Reviews of Physiology, Biochemistry and Pharmacology 150

Springer

Berlin
Heidelberg
New York
Hong Kong
London
Milan
Paris
Tokyo

Reviews of
150 Physiology
Biochemistry and
Pharmacology

Editors

S.G. Amara, Portland • E. Bamberg, Frankfurt
M.P. Blaustein, Baltimore • H. Grunicke, Innsbruck
R. Jahn, Göttingen • W.J. Lederer, Baltimore
A. Miyajima, Tokyo • H. Murer, Zürich
S. Offermanns, Heidelberg • N. Pfanner, Freiburg
G. Schultz, Berlin • M. Schweiger, Berlin

With 32 Figures and 5 Tables

 Springer

ISSN 0303-4240
ISBN 3-540-20214-5 Springer-Verlag Berlin Heidelberg New York

Library of Congress-Catalog-Card Number 74-3674

Springer-Verlag is a part of Springer Science+Business Media

springeronline.com

© Springer-Verlag Berlin Heidelberg 2004
Printed in Germany

Cover design: design & production GmbH, Heidelberg

Printed on acid-free paper – 14/3150 ag   5 4 3 2 1 0

Rev Physiol Biochem Pharmacol (2003) 150:1–35
DOI 10.1007/s10254-003-0018-9

H.-J. Apell

Structure–function relationship in P-type ATPases—a biophysical approach

Published online: 17 June 2003

Abstract P-type ATPases are a large family of membrane proteins that perform active ion transport across biological membranes. In these proteins the energy-providing ATP hydrolysis is coupled to ion-transport that builds up or maintains the electrochemical potential gradients of one or two ion species across the membrane. P-type ATPases are found in virtually all eukaryotic cells and also in bacteria, and they are transporters of a broad variety of ions. So far, a crystal structure with atomic resolution is available only for one species, the SR Ca-ATPase. However, biochemical and biophysical studies provide an abundance of details on the function of this class of ion pumps. The aim of this review is to summarize the results of preferentially biophysical investigations of the three best-studied ion pumps, the Na,K-ATPase, the gastric H,K-ATPase, and the SR Ca-ATPase, and to compare functional properties to recent structural insights with the aim of contributing to the understanding of their structure–function relationship.

Introduction

All living cells are surrounded by membranes that separate their strictly controlled cytoplasmic contents from their environment, and within cells numerous compartments with specific functions and different compositions of components are enclosed also by membranes. These membranes consist of lipid bilayers, which are effective barriers for most of the water-soluble substances, such as ions, sugars, and amino acids. To perform its metabolism, a cell needs selective and controlled transport of substrates and of end products of the metabolic processes across these membranes. This transport function is performed by membrane proteins.

H.-J. Apell (✉)
Department of Biology,
University of Konstanz,
Fach M635, 78457 Konstanz, Germany
e-mail: h-j.apell@uni-konstanz.de

Besides the separation of aqueous phases, a second function of membranes is the storage of energy in the form of chemical potential gradients, $\Delta\mu_i = RT \cdot \ln \left(c_i'/c_i''\right)$, of substances i in the case of uncharged substances or in the case of ions in the form of electrochemical potential gradients, $\Delta\tilde{\mu}_i = RT \cdot \ln \left(c_i'/c_i''\right) + z_i F(\varphi' - \varphi'')$.

On the basis of thermodynamic principles, two classes of transport proteins can be discriminated: proteins that perform passive and active transport. Passive transport is defined by facilitated diffusion "downhill" along the (electro-)chemical potential gradient of the transported substance whereby the energy gradient dissipates. Active transport occurs "uphill," increasing the (electro-) chemical potential of the transported substances. This is possible only if energy in the form of free energy, ΔG, is provided from another process which is coupled to the transport across the membrane. This energy input has to be larger than the (electro-) chemical potential, $|\Delta G|gt; \Delta\tilde{\mu}_i$. Active ion-transport proteins in animals are mostly ion transporters, so-called ion pumps. A careful and detailed introduction into the biophysics of ion pumps can be found in the monograph *Electrogenic Ion Pumps* (Läuger 1991).

Energy sources that power active ion transport are light, e.g., in bacteriorhodopsin (Stoeckenius 1999; Der and Keszthelyi 2001; Lanyi and Luecke 2001), redox energy, e.g., in the cytochrome c oxidase (Michel 1999; Wikstrom 2000; Abramson et al. 2001), or decarboxylation, e.g., in ion-translocating decarboxylases (Dimroth 1987; Michel 1999; Wikstrom 2000; Abramson et al. 2001). The most common energy-producing mechanism is, however, ATP hydrolysis in transport ATPases.

Ion-motive ATPases are the largest and most diverse class of ion pumps. Three groups are discussed in the literature: (a) F-type ATPases (Dimroth et al. 2000; Papa et al. 2000; Capaldi and Aggeler 2002; Senior et al. 2002), which work in many cases in reverse direction as so-called ATP synthetases, e.g., in the inner mitochondrial membrane or in the thylakoid membrane of chloroplasts. (b) V-type ATPases (Sze et al. 1992; Nelson 1995; Forgac 1999) which are ubiquitous H-ATPases with a structure related to that of F-type ATPases. They are found in cellular organelles of an ever-increasing number of different cells. (c) P-type ATPases, which are found in virtually all eukaryotic cells and also in bacteria.

P-Type ATPases

In contrast to the other two types of ion-motive ATPases, P-type ATPases are of a much simpler structure (Møller et al. 1996). They have an α-subunit of approximately 100 kDa that contains all components essential for enzymatic activity and transport. Examples of such single-subunit P-type ATPases are, e.g., Ca-ATPases (Carafoli 1992; Lee and East 2001). Na,K-ATPase and H,K-ATPase are functional only if assembled together with a β-subunit (McDonough et al. 1990; Geering et al. 2000; Geering 2001). In the case of the Na,K-ATPase, in specific tissues a γ-subunit was found, which is also discussed as a regulatory device (Berrebi-Bertran et al. 2001; Cornelius et al. 2001; Therien et al. 2001). Meanwhile, a whole family of such regulators was identified, called FXYD proteins (Beguin et al. 2002; Garty et al. 2002). A K-ATPase of *E. coli* (the so-called Kdp-ATPase) is composed of three different polypeptides (Epstein et al. 1990; Altendorf et al. 1992).

A second fundamental difference between P-type ATPases and the other ion-motive ATPases is their enzymatic reaction mechanism (Glynn 1985; Lancaster 2002), which

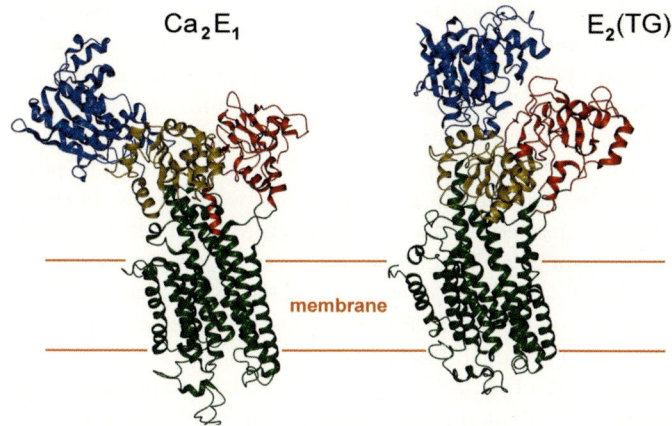

Fig. 1 Structure of the Ca-ATPase of the sarcoplasmatic reticulum in both principal conformations as resolved by their crystal structure. *Left:* In its conformation Ca_2E_1 (PDB file 1EUL) the spatial resolution was 2.6 Å (Toyoshima et al. 2000). *Right:* The structure in the E_2 conformation (PDB File 1IWO) was stabilized by tharpsigargin (not shown) and obtained from crystals with a resolution of 3.1 Å. (Toyoshima and Nomura 2002)

contains a phosphorylated intermediate. The γ phosphate of ATP is transferred to a highly conserved aspartyl residue in the large cytoplasmic loop between the forth and fifth transmembrane segment. Specific to P-type ATPases is also that the enzymatic activity (and consequently ion transport) can be inhibited by ortho-vanadate, which acts as a tightly bound transition-state analogue of phosphate (Cantley et al. 1977).

P-type ATPases are found in virtually all eukaryotic cells and also in bacteria, where they actively transport various ions. They are distributed in different classes (I–IV) and several subgroups (Sweadner and Donnet 2001) according to the ions they transport: Na^+, K^+, Ca^{2+}, H^+, Mg^{2+}, Cu^{2+}, Cd^+, Hg^+, and even Cl^- (Gerencser 1996).

Structural properties

Although their molar masses vary between about 70 and 100 kDa, the first five transmembrane domains and the large cytoplasmic loop, which forms the main part of the enzymatic machinery, are well conserved for all P-type ATPases. Yeast proteins mostly have six transmembrane domains, while those from animal tissues preferentially have ten (Sweadner and Donnet 2001).

A breakthrough in the understanding of structure–function relationships was made when the first highly-resolved 3D structure of a P-type ATPase became available with a resolution of 2.6 Å (Fig. 1), the Ca-ATPase of the sarcoplasmatic reticulum in its E_1 conformation with 2 Ca^{2+} ions bound ("Ca_2E_1;" Toyoshima et al. 2000). The structure confirms the topological organization of ten transmembrane helices deduced for Ca, Na,K-, H,K- and H-pumps by biochemical techniques (MacLennan et al. 1985), and the structure reveals several unexpected features. It was found (Toyoshima et al. 2000) that (a) both ions are located side by side with a distance of 5.7 Å close to the middle of the transmembrane section of the protein, (b) the ion binding sites are surrounded by the transmembrane helices M4–M6 and M8, (c) the α helices M4 and M6 are partly unwound to provide an efficient coordination geometry for the two Ca^{2+} ions, and (d) a cavity with a rather wide opening, surrounded by M2, M4, and M6 is discussed as an access structure on the cytoplasmic side. The outlet of Ca^{2+} is likely to be located in the area surrounded by M3–M5. The details of Ca^{2+} occlusion sites fit well with that deduced in extensive mutagenesis studies (Clarke et al. 1989a, MacLennan et al. 1997). The parts of the protein protruding into the cytoplasm are divided into three domains, two domains, N (nucleotide) and P (phosphorylation), are formed by the loop between M4 and M5, well separated from a third A domain (actuator or anchor) formed by the loop between M2 and M3 and the tail leading into M1. The fold of the P-domain is like that of L-2-haloacid dehalogenase and related proteins with homologies to P-type pumps in conserved cytoplasmic sequences (Saraste et al. 1990; Aravind et al. 1998).

Recently, the structure of the SR Ca-ATPase in its second principle conformation, E_2, stabilized by the specific inhibitor tharpsigargin ["E_2(TG)"], became available with a resolution of 3.1 Å (Toyoshima and Nomura 2002). Due to the low Ca^{2+}-binding affinity in the E_2 state, it was not possible to obtain crystals with Ca^{2+} ions bound which would allow a direct determination of the position of the binding sites. It was proposed that in E_2(TG) the counterions H^+ are bound and access to the luminal sites is already locked. Nevertheless, by comparison of both crystallized forms, Ca_2E_1 and E_2(TG), it is possible to describe a number of changes in the protein structure that are important for conclusions on functional properties related to enzymatic and transport activity. [These differences are impressively visualized as supplementary information to Toyoshima and Nomura (2002) on *Nature*'s website (www.nature.com).] The three cytoplasmic domains, N, P, and A, which form the enzymatic machinery, are wide open in the Ca_2E_1 form, and they are folded together to a much more compact assembly in the E_2(TG) form (Fig. 1). This transition requires movements of the N domain of about 50 Å and a rotation of the A domain of about 110°. The cytoplasmic domains move as a whole in a M10-to-M1 direction (Toyoshima and Nomura 2002). The P and N domains themselves are not changed between both conformations.

Despite the previously often discussed concept that in the membrane domains no major structural rearrangements are expected between different conformations of the pump, the reported changes of position and tilt of the first six transmembrane helices are dramatic. The transition between Ca_2E_1 and E_2(TG) is rather complicated and includes partial unwinding of α-helices, bending a part of an α-helix by almost 90° (M1), changing tilts (M2-M5), ~90° rotations (M6), shifts towards the cytoplasmic side (M1, M2) or shifts in opposite direction by 5 Å, which is almost one turn of an α-helix (M3, M4). (For more details see Toyoshima and Nomura 2002.) With respect to the cytoplasmic domains of the enzymatic machinery the interplay between these is obvious, and it is clearly possible to imagine the concept that Ca^{2+} binding has to trigger enzyme phosphorylation, and that a relaxation of the phosphorylated form (and release of the nucleotide) subsequently disrupts the ion binding sites as seen in the crystallized E_2(TG) conformation. The almost perfect coordination of both Ca^{2+} ions in E_1 (Toyoshima et al. 2000) is abolished in E_2(TG) by a shift of M4 and a clockwise rotation of the three crucial residues on M6 out of site I (Toyoshima and Nomura 2002).

So far, the SR Ca-ATPase is the only P-type ATPase with such a detailed structural resolution. From other members of this family only images with a lower resolution of about 8 Å are available (Kühlbrandt et al. 1998; Scarborough 1999; Hebert et al. 2000). A com-

Fig. 2 Reaction scheme for a P-type H-ATPase. E_1 and E_2 are the conformations of the protein with ion-binding sites facing cytoplasm and extracellular medium, respectively. Certain transitions between neighboring states of the protein must be kinetically inhibited (*dashed lines*) to produce ATP-driven transport cycle (*solid lines*) that pumps H^+ ions out of the cell

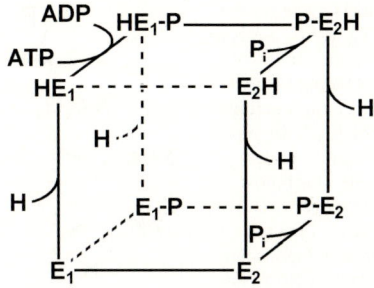

parison of such images with a similarly resolved SR Ca-ATPase structure (Zhang et al. 1998) indicates that they agree in most of the important structural details. Therefore, a computer approach was used in which the conserved homology, especially in the ATP hydrolysis site of the P-type ATPases (Jørgensen et al. 2001), as well as other aligned conserved segments, were mapped on the SR Ca-ATPase structure. Although the homology is highest for the Na,K-ATPase, this procedure led to reasonable results also for other P-type ATPases (Sweadner and Donnet 2001). Most insertions and deletions were predicted to be at the protein surfaces, and the similarity proposes a shared folding of all tested P-type ATPases, despite some particular exceptions.

Therefore, the structural features of the SR Ca-ATPase will be used in the following paragraphs to represent the considerations of structure–function principles of P-type ATPases.

Principles of transport functions

The eminent importance of the insights into structural details of the SR Ca-ATPase is paired with a functional analysis, which is most elaborate for the Na,K-ATPase. The transport mechanism found for this ion pump could be generalized for all P-type ATPases. Since enzymatic and transport functions have to be coupled, the pump mechanism has to be a complex process.

The analysis of the ion-transport process in P-type ATPases revealed that at least three categories of reactions have to occur, performed sequentially in forward or backward direction: (a) ion binding or release, (b) ion occlusion or deocclusion, and (c) transitions between both principal conformations in which the binding sites become accessible from the cytoplasm (E_1) or from the opposite aqueous compartment (E_2). Taking these reactions into account, a general reaction scheme can be constructed which has eight states in the simplest case of a H-ATPase that transfers one H^+ per hydrolyzed ATP (Fig. 2). If all transitions were allowed, such a protein would short-circuit the membrane for H^+ ions in the fashion of an ion carrier, and it would be able to dissipate the energy provided by ATP hydrolysis without ion transport. Therefore, a number of transitions have to be inhibited kinetically by the pump protein to perform active ion transport as indicated in Fig. 2 by dashed lines (Läuger 1991). Indeed, this reaction scheme was found to represent perfectly the function of a P-type H-ATPase from *Enterococcus hirae* (Apell and Solioz 1990). The transport of counterions, as found in most of the other P-type ATPases, can be constructed

Rev Physiol Biochem Pharmacol (2003) 150:1–35

Fig. 3 Post–Albers cycle for the Na,K-ATPase. E_1 and E_2 are conformations of the ion pump with ion binding sites facing the cytoplasm and extracellular medium, respectively. $(Na_3)E_1$-P, $E_2(K_2)$ and $E_2(K_2)\cdot$ATP are occluded states in which the ions bound are unable to exchange with either aqueous phases. Enzyme phosphorylation and dephosphorylation occurs on the cytoplasmic side of the protein

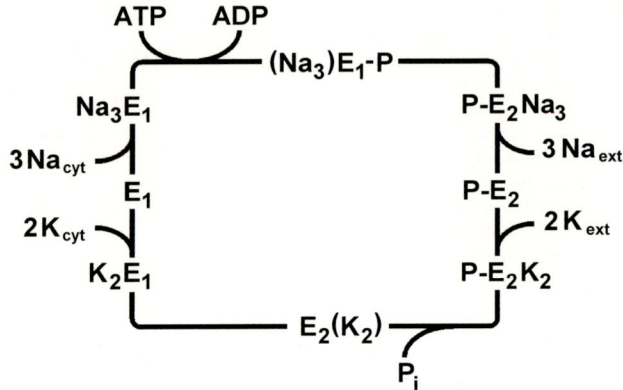

from such a simple scheme by stacking a second, analogous reaction cube underneath the one shown.

The Na,K-ATPase maintains the electrochemical potential gradient of Na^+ and K^+ ions across cell membranes (Läuger 1991; Apell 1997; Glitsch 2001; Jørgensen and Pedersen 2001). The ion transport is facilitated by coupling the energy-providing enzymatic process with a Ping-Pong mechanism of ion translocation. This process is described by the so-called Post–Albers cycle (Albers 1967; Post et al. 1972) which is in full agreement with the general scheme of Fig. 2. A representation of the Na,K-ATPase pump cycle is shown in Fig. 3. As can be seen from Fig. 3, the reaction sequence of ion binding, occlusion, conformation transition and ion release is performed for both transported ion species. In addition, coupling of the scalar enzymatic activity with the vectorial ion transport takes place in the occlusion reactions: enzyme phosphorylation by ATP together with Na^+ ion occlusion, and enzyme dephosphorylation together with K^+ ion occlusion. Under physiological conditions (i.e., at ATP concentrations above 50 µM), an additional reaction step was found: the so-called low-affinity ATP binding in state $E_2(K_2)$. This reaction, $E_2(K_2)$ + ATP → $E_2(K_2)\cdot$ATP, is not necessary to transport K^+ ions; however, it speeds up the subsequent conformation transition, $E_2(K_2)\cdot$ATP → $K_2E_1\cdot$ATP, by a factor of ten (Simons 1974; Glynn 1985).

Respective "Post–Albers"-type pump cycles were also found for the SR Ca-ATPase (de Meis 1985; Inesi and de Meis 1989), the Ca-ATPase of the plasma membrane (Schatzmann 1989), and the gastric H,K-ATPase (Faller et al. 1985; Helmich-de Jong et al. 1987; Sachs et al. 1992).

Significant differences were observed with respect to the stoichiometry of various P-type ATPases. For Na,K-ATPase, gastric H,K-ATPase and SR Ca-ATPase counter-transport of ions was demonstrated. In the case of the Ca-ATPase, it was not so easy to verify counter-transport since the SR membrane is very leaky for monovalent cations so that neither electric membrane potentials nor pH gradients can be monitored reliably across the SR membrane. Only after reconstitution of the SR Ca-ATPase in proteoliposomes convincing evidence was produced that it is a Ca,H-ATPase (Yu et al. 1993, 1994). For a number of other ATPases, studies of the transport properties are not sufficiently advanced to establish counter-transport or stoichiometry.

The three best investigated P-type ATPases showed different stoichiometries: 3 Na^+/2 K^+/1 ATP for the Na,K-ATPase, 2 H^+/2 K^+/1 ATP for the gastric H,K-ATPase, and 2

Fig. 4 Profile of the potential energy of the proton along its pathway in states HE_1, $(H)E_1$-P and P-E_2H. The high energy barriers symbolize a virtually impenetrable structure for an H^+ ion. α', β', β'' and α'' represent relative dielectric distances which characterize the fraction of the membrane potential that has to be traversed by ions between two neighboring pump states. In the occluded state, equilibration between binding site and aqueous phase is blocked on both sides. Nonzero values of the dielectric distances correspond to an electrogenic contribution. Dielectric and spatial distances are not necessarily the same

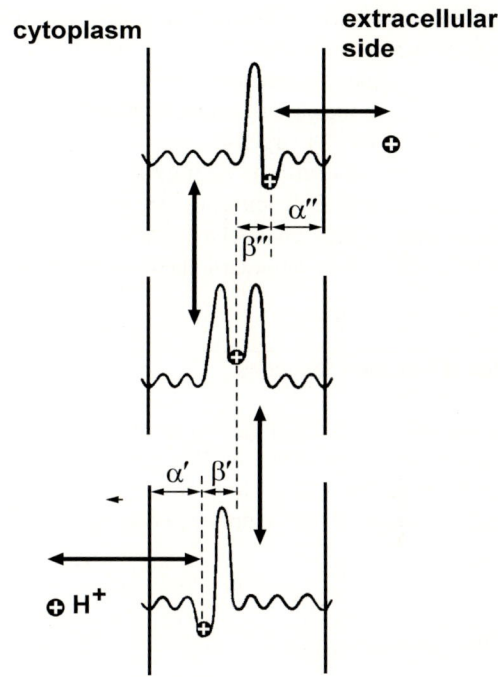

Ca^{2+}/2 H$^+$/1 ATP for the SR Ca-ATPase. From the amount of charges transported per cycle, an important parameter is determined—the electrogenicity of the pump. It is defined as the number of elementary charges moved out of the cytoplasm per molecule ATP hydrolyzed. From the numbers given above, the following electrogenicities are obtained: +2 (SR Ca-ATPase), +1 (Na,K-ATPase) and 0 (gastric H,K-ATPase). However, even when no net charge is translocated across the membrane, as in the case of the H,K-ATPase, ions are moved through the membrane (or protein) dielectric. Therefore, while proceeding through each half cycle of the respective Post–Albers scheme, charge movements have to occur. An important question is: which of the four steps of a half cycle, (a) ion binding, (b) ion occlusion, (c) conformation transition, and (d) ion release to the opposite side, is associated with a shift of charge(s) within the membrane dielectric, and how large is its contribution to the total charge movement?

To quantify such charge movements we consider a potential-energy profile of an ion along its transport pathway in different states of a half cycle (Fig. 4). For the sake of simplicity, again an H-ATPase is chosen which transports one H^+. The "dielectric" distance (or "dielectric coefficient") was introduced as a characteristic parameter to describe the fraction of membrane dielectric over which the charge is shifted perpendicular to the plane of the membrane (Läuger 1991). If a dielectric coefficient is nonzero, the respective reaction step in the Post–Albers scheme is termed "electrogenic." In Fig. 4, for example, the dielectric coefficient for cytoplasmic binding of a H^+ ion would be α'. According to the conservation principle, the conservation condition, $\alpha'+\beta'+\beta''+\alpha''=1$, has to be fulfilled for the transfer of each ion across the whole membrane.

When the ion-binding sites are inside the membrane part of the protein, as was shown for the SR Ca-ATPase (Toyoshima et al. 2000), for the access to these sites two different limiting cases can be considered: The access may take the form of a wide, water-filled funnel (or vestibule) or a narrow channel which even may require (partial) dehydration of the ions or which may be selective for the transported ion species. The difference between both cases is that in the case of a vestibule, water and all kinds of ions can enter so that the electrical conductance is large, no electrical field can build up, and the drop of the trans-membrane electric potential is negligible. In the other case, part of the transmembrane electric potential will drop across the length of the channel and, therefore, generate a so-called high-field access channel or "ion well" (Mitchell and Moyle 1974; Läuger and Apell 1986). Only in the latter case the movement of ions is electrogenic.

Detection of transport functions

Various experimental techniques were developed and applied to study and analyze trans-port functions of P-type ATPases. The most comprehensive investigations were performed with the Na,K-ATPase. Therefore, in the following paragraphs the various presented ex-perimental techniques will mainly refer to publications on transport properties of this ion pump.

Tracer flux studies

The counter-transport of Na^+ and K^+ allows detailed studies of ion movements in both di-rections through the membrane. Since no K^+ isotopes with convenient half-life times, $t_{1/2}$, are available, ^{86}Rb was used instead. It has a $t_{1/2}$ of 18.7 days, and Rb^+ is a congener of K^+ with similarly high binding affinity (Karlish et al. 1978; Beaugé and Glynn 1979; Glynn and Richards 1982; Schneeberger and Apell 2001). Transport studies were performed in compartmentalized preparations, such as erythrocytes (Sen and Post 1964; Karlish and Glynn 1974; Sachs 1977; Beauge and Glynn 1978), inside-out vesicles of red blood cells (Blostein 1979, 1983), and reconstituted proteoliposomes (Anner et al. 1977; Anner and Moosmayer 1981; Karlish and Pick 1981; Karlish et al. 1982; Cornelius 1991). A rapid filtration method for time-resolved measurements of isotope flux from membrane vesicles was introduced by Forbush with a time resolution of about 30 ms (Forbush, III 1984a, 1984b, 1987). Isotope-flux experiments allow the determination of stoichiometries, ion oc-clusion, and transmembrane movements. However, due to the limited time resolution, ki-netical analyses are possible only for slow processes. In the case of the SR Ca-ATPase, radioactive Ca isotopes were used to demonstrate sequential binding of the two Ca^{2+} ions and transport in the so-called single-file mode (Inesi 1987).

Electrophysiological approaches

A second and by far wider approach to study transport functions and kinetics are electro-physiological methods. In this field especially the Na,K-ATPase was scrutinized. The Ca-ATPase in SR membranes cannot be measured by direct electric techniques since the (leak) conductance of this membrane is high due to the permeability for monovalent ions. The field of methodological approaches for the Na,K-ATPase reaches from investigation of whole epithelia (Horisberger and Giebisch 1989) to squid axon (Rakowski et al. 1987), cardiac myocytes (Gadsby et al. 1985; Nakao and Gadsby 1986), Purkinje cells (Glitsch and Tappe 1995) and giant membrane patches (Hilgemann 1994). *Neurospora* cells were used to study a H-ATPase (Slayman and Sanders 1985). Injection of mRNA into oocytes became a very successful technique, not only to investigate in a convenient way different isoforms of the Na,K-ATPase, but also to study mutated pumps or chimera between Na,K-ATPase and H,K-ATPase or SR Ca-ATPase (Lafaire and Schwarz 1985; Horisberger et al. 1991; Jaunin et al. 1993; Zhao et al. 1997; Mense et al. 2000). Additional access to the Na,K-ATPase expressed in oocytes was obtained by the cut-open technique that allows internal perfusion of the oocytes in a simple way (Holmgren and Rakowski 1994). In such cellular membrane systems both sides of the membrane are accessible separately and besides transmembrane pump currents, the current-voltage dependencies of the ion pumps can be determined. These so-called I-V curves provide important information on the pump mechanism (Läuger and Apell 1988; Läuger 1991; de Weer et al. 2000). In experiments with native membranes, a prominent problem is the presence of various other ion-transport systems, such as ion channels and ion carriers, which may produce electric currents larger than that from ion pumps. Therefore, the other ion-transport proteins have to be blocked by inhibiting agents, and the residual currents have to be measured in the absence and presence of specific pump inhibitors. The difference of the currents in the absence and presence of the inhibitor represents the pump-specific ion transport. Certain cardiac glycosides are appropriate inhibitors for those experiments in the case of the Na,K-ATPase (Lederer and Nelson 1984; Gadsby 1984).

To overcome the difficulties of native membranes, a supplementary approach was chosen in which purified membrane preparations were used so that the only remaining transport protein was the Na,K-ATPase (Jørgensen 1974). Since the resulting membranes are no longer vesicular but flat membrane patches which have sizes in the order of 1 µm diameter or less (although they have densities of up to 7,000 pump molecules per μm^2), their transport properties cannot be measured directly with electrodes on both sides of the membrane. However, they became accessible to electric studies on the basis of a proposal by Peter Läuger who suggested adsorbing the Na,K-ATPase-containing membranes onto black lipid membranes (BLM) and triggering the pump action with an ATP-concentration jump by release of ATP from its inactive precursor, caged ATP (Fendler et al. 1985; Borlinghaus et al. 1987; Apell et al. 1987; Fendler et al. 1988, 1993; Sokolov et al. 1998). This method was also applied in studies of the SR Ca-ATPase (Hartung et al. 1987) and of the gastric H,K-ATPase (Fendler et al. 1988). A more recent development in this technique is the use of so-called solid-supported membranes, which are much more stable than BLM and which allow an easy exchange of the buffer composition (Seifert et al. 1993; Pintschovius and Fendler 1999; Domaszewicz and Apell 1999).

Reconstitution of Na,K-ATPase in BLMs turned out to be tricky since the ions pumps tend to denature and to form ion-channel-like structures during this procedure (Reinhardt et al. 1984). The results with incorporated, active ion pumps were not easily reproducible and generated rather small currents under turnover conditions (Eisenrauch et al. 1991). Similar small currents (<30 fA) were reported for reconstituted SR Ca-ATPase (Eisenrauch and Bamberg 1990; Nishie et al. 1990).

Measurements of the electric current through the Na,K-ATPase can be combined with tracer flux experiments to analyze transport stoichiometry in squid axon (de Weer et al. 1988, 2001) and in oocytes (Schwarz and Gu 1988; Rakowski 1989).

Fig. 5 pH and conformation dependence of the 5-IAF label co-
valently linked to rabbit $\alpha 1$
Na,K-ATPase. The conforma-
tion-dependent shift of the titration curve fitted through the experimental data indicates that the
local pH in the environment of
the label is modified by rearrangements of amino-acid side
chains. The respective pK values
of the curves are 6.55 (Na$_3$E$_1$),
6.7 (E$_1$) and 6.8 (E$_2$(K$_2$)). The
conformation-induced shift explains the fluorescence changes
when the detection is performed
at constant bulk pH

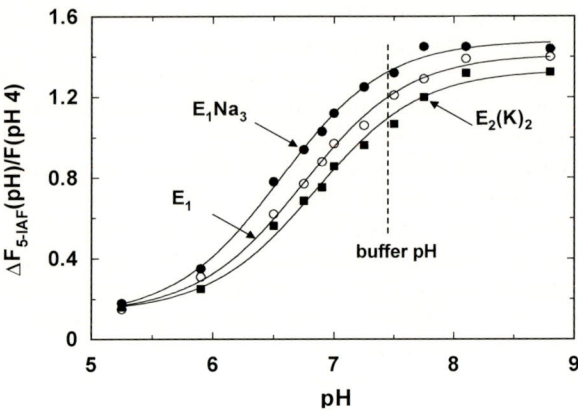

Fluorescence methods

Over the past two decades fluorescence methods were advanced to gain detailed information on structural changes and transport properties of P-type ATPases. Intrinsic tryptophan
fluorescence (Karlish and Yates 1978; Boldyrev et al. 1983; Demchenko et al. 1993; Møller et al. 1996; Ferreira and Coelho-Sampaio 1996) as well as the fluorescence of covalent-
ly bound labels or of membrane soluble dyes ("extrinsic fluorescence") can be used to detect function-dependent responses of the proteins.

The first set of applied labels were fluoresceine derivatives which could be bound cova-
lently to specific amino acids and which reported conformational changes. Fluorescein-5-
isothiocyanate (FITC) binds to Lys-501 within the ATP-binding site and thus prevents
ATP binding and enzyme phosphorylation by ATP (Hegyvary and Post 1971; Sen et al.
1981; Farley and Faller 1985). However, FITC responds with a significant fluorescence
change to the conformation transition between E$_1$ and E$_2$ (Rephaeli et al. 1986; Karlish
1988). Recently it was demonstrated that FITC also reports binding of the third Na$^+$ ion to
the Na,K-ATPase (Schneeberger and Apell 1999). Due to the high conservation of the
ATP-binding site in the P-type ATPases, FITC binding could be performed successfully
also with SR Ca-ATPase (Kirley et al. 1985; Seidler et al. 1989) and with gastric H,K-
ATPase (Asano et al. 1989; Faller et al. 1991).

The second fluoresceine derivative that detects conformation transitions of the Na,K-
ATPase is 5-iodoacetamidofluorescein (5-IAF) (Kapakos and Steinberg 1982, 1986). This
label binds to Cys-457 (Tyson et al. 1989), well away from the ATP binding site, so that
the protein can be phosphorylated by ATP and is able to perform its complete pump cycle,
a clear advantage over the FITC label. However, for the $\alpha 1$ isoforms of the Na,K-ATPase
from various animals, the substrate-induced fluorescence changes of 5-IAF showed significant differences: no responses were found in pig enzyme, intermediate responses in rabbit
enzyme, and maximal responses in dog enzyme (Steinberg and Karlish 1989; Stürmer et
al. 1989). The underlying mechanism of these fluoresceine labels is that of a pH indicator
which responds to small pH changes in the local environment of the label. In Fig. 5, pH
titrations of the 5-IAF fluorescence intensity are shown for three different states of labeled
rabbit kidney Na,K-ATPase. It can be seen that changes of the protein conformation, induced by additions of substrates such as Na$^+$, ATP, and K$^+$, shift the titration curves, probably due to small variations of the protein-surface shape near the attached fluorescent label

and, in consequence, charged amino-acid side chains affect the local ion concentration on the protein surface (by a Guy-Chapman effect) and, consequently, also the local pH. When the bulk pH is buffered, as it was in the experiments performed to study the pump action, the fluorescence intensity is modulated with protein-conformation changes.

Further fluorescent labels applied to monitor conformational changes are eosin (Skou and Esmann 1981, 1983; Lin et al. 1997) and erythrosin 5'-isothiocyanate (Linnertz et al. 1998a, 1998b). Various fluorescent labels which can be bound simultaneously at different locations of the pump protein can also be used to determine distances between selected domains by the Förster resonance energy transfer mechanism (Linnertz et al. 1998b).

Fluorescent probes were also used to detect ion transport by P-type ATPases. In such studies, ion pumps are reconstituted in lipid vesicles, in which the inside-out oriented AT-Pases can be activated by addition of ATP, and transport activity is monitored by a fluorescence response to the generated electric membrane potential due to the electrogenicity of the ion pump. For a quantitative analysis, inhomogeneities of the vesicles in diameter and number of active pump molecules have to be taken into account (Apell and Läuger 1986). Appropriate dyes are 1,3,3,1',3',3'-hexamethylindodicarbocyanine (NK529; Apell et al. 1985), oxonol VI (Apell and Bersch 1987) and the carbocyanine dye DiS-C3-(5) (Goldshlegger et al. 1987). These dyes redistribute between aqueous phase and membrane as a function of the membrane potential across the membrane. Therefore, their time resolution is limited by the redistribution process. Typical time constants are on the order of 300 ms (Clarke and Apell 1989). With these assays, numerous transport properties could be determined for the Na,K-ATPase (Cornelius 1989; Clarke et al. 1989b, Goldshleger et al. 1990; Apell et al. 1990), for the SR Ca-ATPase (Cornelius and Møller 1991; Yu et al. 1993, 1994), and for a H-ATPase from *Enterococcus hirae* (Apell and Solioz 1990). The H,K-ATPase cannot be studied with such an approach due to its overall electroneutrality.

Styryl dyes, such as RH160, RH237, and RH421, were used since 1988 to trace pump activity of the Na,K-ATPase (Klodos and Forbush, III 1988; Bühler et al. 1991). These dyes, and others of this family, are hydrophobic compounds of amphiphilic character, which insert into lipid membranes in an aligned manner (Pedersen et al. 2001). Due to their electrochromic mechanism, they detect changes of local electric fields in the membrane dielectric (Loew et al. 1979; Fluhler et al. 1985) and, therefore, report charge movements in membrane preparations in which ion pumps are present in a sufficiently high density ($>10^3/\mu m^2$; Pedersen et al. 2001). Styryl dyes are so-called fast dyes since their response times are in a submicrosecond range. They can be applied to membrane vesicles or to open membranes, such as purified microsomal preparations of the Na,K-ATPase, so that a transmembrane ion transport is not detected but movements of ions into the membrane domains of the ion pumps or their release into the aqueous phase are detected (Stürmer et al. 1991). (RH421 can also be used to follow the transmembrane potential generated by Na,K-ATPase action; however, typical fluorescence changes were 15% per 100 mV transmembrane potential, while oxonol VI showed about 100% change in the same experiment.) With respect to the Na,K-ATPase, a variety of styryl dyes were tested which produced differently large responses for the electrogenic partial reactions (Bühler et al. 1991; Fedosova et al. 1995; Pedersen et al. 2001). With this method, a wide spectrum of partial reactions of the Na,K-ATPase was studied and analyzed which resulted in an advanced understanding of ion binding affinities (Stürmer et al. 1991; Bühler and Apell 1995; Schneeberger and Apell 1999, 2001), rate constants of single reaction steps (Pratap and

Robinson 1993; Heyse et al. 1994; Visser et al. 1995; Clarke et al. 1998; Humphrey et al. 2002) and the energetics of the pump cycle (Apell 1997).

Styryl dyes were also used to investigate an H-ATPase from *Neurospora* (Nagel et al. 1991), the SR Ca-ATPase (Butscher et al. 1999; Peinelt and Apell 2002), and recently the possibility of a comparable application with the gastric H,K-ATPase was demonstrated (Diller et al. 2003).

Inhibitors

All P-type ATPases may be functionally blocked by inhibitors. A common inhibitor of all P-type ATPases is ortho-vanadate, which binds with significantly higher affinity than phosphate to the phosphorylation site (Cantley et al. 1977; Stankiewicz et al. 1995) and blocks the pumps in their occluded E_2 state. Specific inhibitors for the different ATPases are widely used to discriminate the activity of a single pump species or to "freeze" the protein in a defined conformation.

The Na,K-ATPase has been known for half a century to be inhibited by cardiac glycosides (Schatzmann 1953), a whole family of compounds of which ouabain is the most well known (Glynn 1985). Cardiac glycosides block the Na,K-ATPase from the outside of the membrane.

In the case of the gastric H,K-ATPase, a well-known inhibitor is the compound SCH28080, which also inactivates from the outside of the cell by blocking the access of the binding sides for K^+ ions (Keeling et al. 1988; Vagin et al. 2002).

The most frequently used inhibitor of the SR Ca-ATPase is tharpsigargin (Lytton et al. 1991; Inesi and Sagara 1992). It blocks the protein in its E_2 conformation and the interaction between inhibitor and protein is known in detail since it could be resolved in the crystal structure of the protein-inhibitor complex (Toyoshima and Nomura 2002).

Detailed ion-transport mechanism

The methods introduced above were used during the last two decades to resolve the pump mechanisms of a number of P-type ATPases. The three ion pumps for which detailed information is available will be discussed in the following paragraphs. Since the number of publications that contain contributions to the mechanism is so large, in the following paragraphs, references will be provided mainly to recent review-type articles, which allow access to the abundance of data, and to a few articles which contain important new insights.

Na,K-ATPase

The Na,K-ATPase is a crucial transport protein of all animal cells which maintains the electro-chemical potentials for Na^+ and K^+ ions across the cytoplasmic membrane at the expense of ATP hydrolysis. The K^+ concentration gradient controls mainly the electric membrane potential, which is reflected by the fact that the electrochemical gradient for K^+ is close to its thermodynamical equilibrium. In contrast, the electrochemical potential for Na^+ is kept far away from its equilibrium and is, therefore, an energy source for many transmembrane processes, such as the initial part of the action potentials in excitable cells and secondary active transport proteins which couple uphill transport of sugars, amino acids, or Ca^{2+} ions to the downhill movement of Na^+. (For a review see Läuger 1991.)

Recent synopses on the investigation of ion transport by the Na,K-ATPase can be found in several reviews (de Weer et al. 2000; Pavlov and Sokolov 2000; Apell and Karlish 2001). As can be seen from the Post–Albers scheme in Fig. 3, the ion transport can be split into a sequence of outward Na^+ transport followed sequentially by an inward K^+ transport. Both branches were analyzed in great detail.

Forward Na^+ transport, $E_1 + 3\ Na^+_{cyt} \rightarrow E_2\text{-}P + 3\ Na^+_{ext}$, requires ATP, and ATP hydrolysis occurs only with 3 Na^+ bound to the protein. Even if only Na^+ ions are present on the cytoplasmic side of the pump, virtually no transition into a state $E_2(Na_2)$ has been found, in contrast to other congener cations (K^+, Rb^+, Cs^+, NH_4^+, Tl^+), which antagonize Na^+ binding and cause a conformational change into the occluded E_2 state after two ions have bound. The virtual absence of the state $E_2(Na_2)$ is in agreement with the observation that binding of the third Na^+ ion occurs with a higher affinity than binding of the second, and stabilizes the protein in the Na_3E_1 state. This can be understood only by assuming that the third Na binding site becomes available after two Na^+ ions have already bound (Schneeberger and Apell 2001). Binding of the first two Na^+ ions was found to be apparently electroneutral (like binding of 2 K^+ ions or their congeners); binding of the third Na^+ ion is electrogenic with a dielectric coefficient of 0.25 (Domaszewicz and Apell 1999). The time resolution of the techniques available to study cytoplasmic Na^+ binding is not yet high enough to determine the rate constant of these reaction steps; only equilibrium dissociation constants could be obtained. Occupation of the third, highly selective Na^+ binding site, which is also formed by transmembrane parts of the Na,K-ATPase, is strictly correlated with a detectable effect on the fluorescent FITC-labeled enzyme. This is interpreted as an Na^+-induced structural transition in the nucleotide binding site, probably a transition state between the "open" configuration of the N, P, and A domains of the cytoplasmic part of the protein to a more compact or "closed" one as observed in the E_2 conformation, as can be seen in Fig. 1 (Toyoshima and Nomura 2002). This transition includes a movement of bound ATP into a position where its γ phosphate becomes able to coordinate with Asp-371, the phosphorylation site. Thus, binding of the third Na^+ ion enables the enzyme to become phosphorylated, and this "trigger" ensures that no ATP is wasted unless three Na^+ ions are bound inside the pump. The subsequent phosphorylation of the enzyme is correlated with an occlusion of the three Na^+ ions, $Na_3E_1\cdot ATP \rightarrow (Na_3)E_1\text{-}P + ADP$. This process is electroneutral, i.e., no net charge movement within the membrane domain could be detected (Borlinghaus et al. 1987). Therefore, the rate constants of this step could be determined only indirectly. Assuming that the enzymatic and transport-coupled reaction are tightly correlated, the rate constants of enzyme phosphorylation, obtained by experiments with radioactive ATP, can be accepted as reference value. The phosphorylation-induced occluded state, $(Na_3)E_1\text{-}P$, is only transient; it cannot be stabilized (unless the protein is treated by oligomycin).

When phosphorylated by ATP, the enzyme performs a conformational transition into its $E_2\text{-}P$ states, in which the bound Na^+ ions are successively deoccluded and released. The voltage sensitivity of this partial reaction was demonstrated with internally perfused squid giant axons and the major component of charge movement was assigned to the Na^+ release (or binding) steps (Gadsby et al. 1993). While the conformational relaxation is of minor electrogenicity, the release of the first Na^+ to the extracellular aqueous phase is the dominant charge-carrying step. It was found that this ion moves through 65%–70% of the pro-

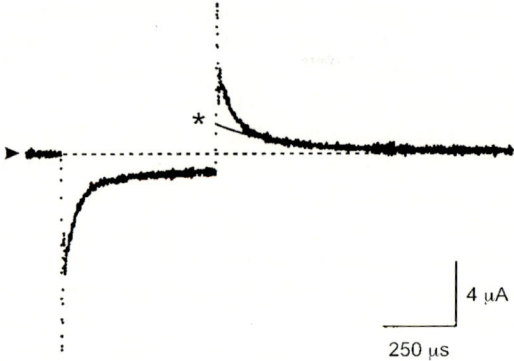

Fig. 6 Electric current caused by extracellular release of Na^+ from the Na,K-ATPase in the membrane of a giant squid axon. The charge movement was elicited by a 500 µs-step from 0 to −110 mV (Holmgren et al. 2000). From such data the presence of three time constants in the range of <30 µs, 250 µs and >4 ms can be derived by fitting with three exponential functions. These processes reflect the Na^+ movements through their "access channel" and intermediate relaxations of the protein structure that deoccluded the binding sites. (From Holmgren at al. 2000, with kind permission)

tein dielectric (Wuddel and Apell 1995). This may be explained by a narrow and deep access channel or "ion well" between the binding site in the protein and the aqueous outside of the protein. The release of the first Na^+ is followed by another conformational relaxation, which brings the remaining two Na^+ ions "electrically" closer to the extracellular aqueous phase because their release in the next reaction steps contributes with dielectric coefficients of 0.1–0.2 only. As shown in Fig. 6, the kinetics of deocclusion/release reactions has been analyzed recently and was found to occur with increasing rate constants from the first ion ($\leqq 1000$ s^{-1}) to the third ($\geqq 10^6$ s^{-1}; Holmgren et al. 2000). The reduced electrogenicity of the second and third Na^+-release step is matched by corresponding dielectric coefficients of K^+ binding (Rakowski et al. 1990), which are the reaction steps following under physiological conditions. Thus, the ion release process may be explained by assuming that the first Na^+ ion moves through a narrow and long access channel while the next two Na^+ ions released face a shallow channel. At least two different mechanisms could explain the transformation from a "deep" to "shallow" ion-well after release of the first Na^+ ion: (a) In a major structural rearrangement of the protein, the narrow ion well widens to become a large vestibule, which then is filled by electrolyte so that the electric-potential surface will come close to the binding sites, or (b) small rearrangements of the α helices allow water molecules to penetrate into the protein matrix from the outside and thus increase the dielectric constant in between binding sites and aqueous phase significantly. This process would also deform the shape of the electric potential within the protein. The latter mechanism would also enable an immediate rehydration of the ions when they are released from their sites without having to migrate as unscreened charges through protein matter whose polarization would be rather energy consuming.

There is convincing experimental evidence that K^+ transport is electroneutral, i.e., that no net charge is moved within the protein between states E_2-$P(K_2)$ and E_1 (Goldshlegger et al. 1987; Rakowski et al. 1990; Domaszewicz and Apell 1999). However, under physiological conditions, the extracellular K^+ concentration is far above the half-saturating concentration of the ion sites so that an electrogenic K^+ binding would be hidden in experiments under this condition (Läuger and Apell 1988). Detailed studies of extracellular K^+ binding steps proved indeed its electrogenicity (Rakowski et al. 1990; Bielen et al. 1991; Bühler and Apell 1995; Peluffo and Berlin 1997).

The partial reaction between E_2-$P(K_2)$ and E_1 can be investigated in the so-called K^+/K^+ exchange mode, in the presence of K^+, with or without Mg^{2+} and with inorganic phosphate, P_i. The experimental findings were explained under the assumptions that (a) the positive charges of the two ions are counter-balanced by two negative charges of the protein, (b) the binding sites in state E_2-P are located inside the protein and are accessible through an ion well, and that (c) in the E_1 conformation binding of K^+ (or its congeners Li^+, Rb^+, Cs^+, Tl^+, and NH_4^+) is electroneutral, i.e., the binding sites are not buried inside the protein (Goldshlegger et al. 1987; Wuddel and Apell 1995; Domaszewicz and Apell 1999).

As will be shown below, in the case of the SR Ca-ATPase and the gastric H,K-ATPase, all ion binding and release steps are electrogenic, an observation which is in agreement with a position of Ca^{2+} ions in the E_1 conformation of the SR Ca-ATPase inside the membrane-spanning parts of the protein (Toyoshima et al. 2000; Toyoshima and Nomura 2002). Therefore, the generally agreed structural similarity of Na,K-ATPase and SR Ca-ATPase (Sweadner and Donnet 2001; Toyoshima and Nomura 2002) argues that ion binding to the Na,K-ATPase ought to be electrogenic, not only in the P-E_2 form, as it was proven, but also in E_1. However, K^+ binding in E_1 was found to be electroneutral, and only binding of the third Na^+ ion appeared to be electrogenic (Domaszewicz and Apell 1999; Schneeberger and Apell 2001). This discrepancy may have two possible explanations: (a) the position of the ion binding sites in the Na,K-ATPase is significantly different from that in the SR Ca-ATPase and H,K-ATPase, or (b) the electrogenicity of the binding and release steps in E_1 is obscured by simultaneous counter-movement of H^+ ions. A hint pointing to the second proposal was found in the fact that in the absence of K^+ and Na^+ it is possible to phosphorylate the enzyme by P_i and that the apparent rates of this pathway are pH-dependent. From the analysis of the kinetics it was concluded that the transition E_2-P $\rightarrow E_1$ occurs with two H^+ ions bound and the transitions with empty binding sites is either extremely slow or absent (Apell et al. 1996). Recently acquired evidence shows that the missing electrogenicity of K^+ release and binding of the first two Na^+ in state E_1 can be explained by an obscuring counter-movement of H^+ ions (Apell and Diller 2002). The obvious ability of the two "non-Na^+ specific" binding sites to bind two H^+ ions in E_1 with an apparent pK that is higher than the cytoplasmic pH under physiological conditions can explain the apparently electroneutral Na^+ and K^+ binding or release. These ion-exchange processes result in apparently electroneutral release and binding steps, and maintain, besides a closely related structural relationship, also a mechanistic agreement between Na,K-ATPase, gastric H,K-ATPase, and SR Ca-ATPase.

Therefore, the biochemically based Post–Albers scheme of the Na,K-ATPase can be expanded as shown in Fig. 7 to explain the transport cycle in greater details. The main difference to the previously proposed pumping mechanism (Apell and Karlish 2001) consists of the placement of all binding sites, as in the case of the SR Ca-ATPase, inside the membrane domains of the protein. Under physiological conditions in conformation E_1 the two binding sites, which are able to bind all kinds of monovalent cations, are always occupied, if not by Na^+ or K^+ ions then by H^+ (Apell and Diller 2002). After two Na^+ ions are bound, the coordination of these ions is assumed to induce a minor conformational rearrangement in the membrane domain, providing access to the third, high-affinity, and Na^+-selective

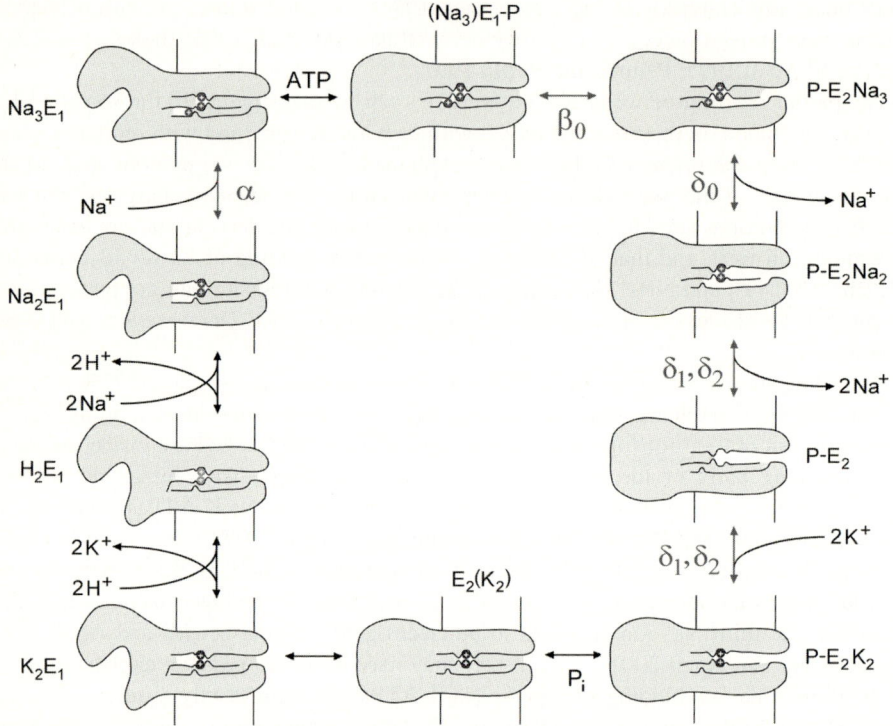

Fig. 7 Refined mechanistic model of the ion transport through the Na,K-ATPase on the basis of the Post–Albers cycle, structural constraints from the Ca-ATPase, and the analysis of charge movements during the transport process. Those reaction steps marked with *Greek letters* indicate the electrogenic processes that were detected under physiological conditions. The corresponding dielectric coefficients are $\alpha=0.25$, $\beta<0.1$, $\delta_0=0.65$, $\delta_1=\delta_2=0.1–0.2$. The apparent electroneutrality (at pH 7) of K^+ release and binding of the first two Na^+ ions on the cytoplasmic side is caused by a counter movement of two H^+ ions

site (Schneeberger and Apell 2001) that does bind an Na^+ (at least under physiological conditions). This step may be accompanied by a so-called preocclusion of the first two Na^+ ions bound, prohibiting their exchange with the aqueous phase. Binding of the third Na^+ to its site about 25% inside the membrane domain is thought to require an additional adaptation of the transmembrane helices to coordinate the ion. This process will affect the N and/or P domain leading to enzyme phosphorylation and the conformation transition into state P-E_2Na_3. As in the case of the Ca-ATPase, it can be expected that this major transition $E_1 \rightarrow E_2$ moves the α-helices of the membrane domain and distorts the coordination of the Na^+ ions in their sites so that the binding affinities decrease by 2–3 orders of magnitude (Wuddel and Apell 1995). This transition shows only a minor electrogenic charge movement within the membrane domain. Due to the small amplitude of this effect, so far it has not been resolved which charges are moved.

Because of the high dielectric coefficient (0.65–0.7), the first Na^+ ion released to the extracellular aqueous phase is assumed to be the ion bound last in E_1. After its removal, a further structural relaxation is proposed (Hilgemann 1994) which allows the remaining two ions to migrate to the aqueous phase with a significantly smaller dielectric coefficient when compared with the first Na^+ ion. An intrusion of a number of water molecules would be sufficient to increase the local dielectric constant sufficiently to account for the reduction in electrogenicity (Wuddel and Apell 1995).

In the P-E_2 state the pK of the binding moieties is significantly lower than in E_1 so that under physiological pH no protonation occurs (Apell and Diller 2002). However, the affinity for K^+ ions is so high in the relaxed P-E_2 state that subsequent K^+ binding occurs spontaneously and fast in an electrogenic manner (Rakowski et al. 1990). Subsequently, occlusion of the K^+ ions due to enzyme dephosphorylation and the conformation transition back into E_1 (with or without ATP in the low-affinity binding site) occur without detectable electrogenicity. The requirement of two negative counter charges in the binding sites to account for the electroneutrality loses weight when the sites remain inside the membrane domain and are no longer shifted close to the cytoplasmic surface, as was required in the previous functional model. On the basis of the Ca-ATPase structure, the primary role of charged amino acids will be creation of an energetically favorable coordination of the dehydrated ion, i.e., of the structure named binding site(s). In E_1 the K^+ ions are able to exchange freely with the aqueous cytoplasm and charged amino acids in the binding sites are immediately compensated by two H^+ ions so that a K^+ release without electrogenic contributions is mimicked.

In summary, the ion-transport mechanism is in agreement with an alternate access model in which the binding sites remain at (almost) the same location and the protein movements open and close the access to these sites alternately on both sides (Läuger 1984).

SR Ca-ATPase

The purpose of the Ca-ATPase of the sarcoplasmatic reticulum is to promote muscle relaxation by pumping Ca^{2+} ions back into the lumen of the reticulum. In this action the protein is able to build up a 10^4-fold concentration gradient across the membrane. With respect to the known stoichiometry of 2 Ca^{2+} ions transported per ATP hydrolyzed, this process is energetically possible only by the fact that no electric potential is generated across the SR membrane. This is made secure in part by counter-transport of 2 H^+ but mainly by a high leak conductance of the membrane for ions other than Ca^{2+}.

As pointed out recently, the snapshots of the SR Ca-ATPase structure in its two basic conformations are a major step in the understanding of pump dynamics of P-type ATPases (Green and MacLennan 2002). Unfortunately, a direct analysis of the transport functions of this ion pump is almost impossible. Due to the mentioned leak conductance of the SR membrane, it is electrically short-circuited, and only processes with time constants short against the RC time of the SR-vesicle membrane may be detected. Therefore, it was difficult to determine the existence of H^+ counter-transport by the pump (Madeira 1978; Chiesi and Inesi 1980), and only after reconstitution of the Ca-ATPase in lipid vesicles (Cornelius and Møller 1991) a proof of counter-transport and of electrogenicity was provided (Yu et al. 1993, 1994). The application of three different fluorescent dyes to detect membrane potential as well as luminal pH and Ca^{2+} concentration demonstrated a stoichiometry of $2Ca^{2+}/2H^+/1ATP$ (Fig. 8). To resolve the electrogenicity of the different reaction steps of the pump cycle, experiments had to be performed with a styryl dye, 2BITC, which showed that Ca^{2+} and H^+ binding and release were accompanied by significant charge movements in the membrane domain (Butscher et al. 1999; Peinelt and Apell 2002). These findings meet the requirements of the position of the ion binding sites of the Ca-ATPase as predicted by the crystal structure of the protein. Kinetical studies of partial reactions in the pump

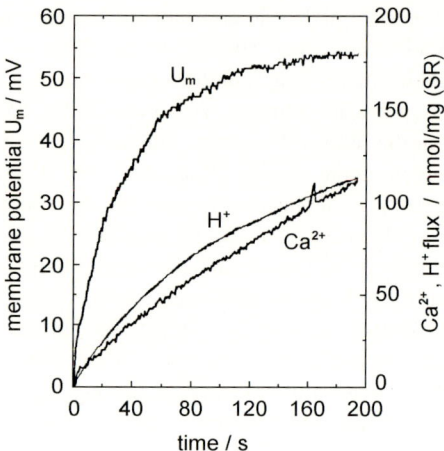

Fig. 8 ATP-dependent Ca^{2+} uptake, H^+ counter transport, and development of transmembrane electrical potential at low temperature. (From Yu et al. 1994, with kind permission). SR Ca-ATPase was reconstituted in lipid vesicles. ATP-induced pump activity was detected with fluorescence dyes: The time course of the membrane potential was detected with oxonol VI, the luminal pH with pyranine, and the luminal Ca^{2+} concentration with arsenazo III. The stoichiometric parallelism of Ca^{2+} uptake and H^+ extrusion is obvious and the voltage increase at low transmembrane electric potentials (when leakage effects are small) is also in agreement with estimations from the amounts of ions transferred

cycle were already performed by analysis of the enzymatic properties many years ago (Andersen and Vilsen 1988; Inesi and de Meis 1989; Inesi et al. 1992). Due to the high leak conductance of the SR membrane, time-resolved measurements are scarce. Studies with radioisotopes and/or quenched flow provided some insight into rate constants for single reaction steps (Froehlich and Heller 1985; Orlowski and Champeil 1991), and capacitive coupling of SR membrane vesicles to a planar lipid bilayer provided some information on the rate-limiting reaction steps in the Ca^{2+} and H^+ transfer (Hartung et al. 1997). A systematic analysis of the kinetical properties beyond early estimations (Inesi and de Meis 1989) is not available so far. Recently, it was shown that the application of styryl dyes may be used similarly successfully to study the time-resolved kinetical behavior of SR Ca-ATPase (Peinelt and Apell 2003). So far, all experimental findings are in agreement with a Post–Albers cycle equivalent to that of the Na,K-ATPase (Fig. 3), in which 3 Na^+ ions are to be replaced by 2 Ca^{2+} and 2 K^+ ions by 2 H^+.

Gastric H,K-ATPase

The gastric H,K-ATPase is enriched in the parietal cells of the gastric glands of the stomach which perform secretion of hydrochloric acid upon hormonal stimulation. The active part in this process is H^+ extrusion from the cytoplasm while, by opening of passive Cl^- and K^+ pathways, both ion species are released, and K^+ is reabsorbed in exchange for the H^+ so that eventually HCl is concentrated up to pH 1–1.5 in the stomach.

The reaction cycle of this P-type pump is also well reproduced by the Post–Albers cycle shown in Fig. 3, in which the Na^+-dependent half-cycle has to be replaced by an H^+-transporting part (with 2 H^+ ions per ATP hydrolyzed). Like in the case of the Na,K-AT-

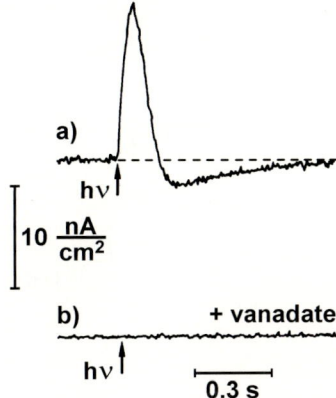

Fig. 9a, b ATP-induced H$^+$ current through gastric H,K-ATPase in membrane vesicles isolated from pig stomach. (From van der Hijden et al. 1990). **a** Buffer contains, besides H$^+$ ions (pH 6), no other monovalent cations. The *arrows* indicate the time when light was switched on to release ~10 μM ATP from caged ATP. The concentration jump triggered the partial reaction H$_2$E$_1$ + ATP → P-E$_2$ + 2 H$^+$ + ADP. The positive current transient represents an inward-oriented flux of positive charge. **b** Inhibition of the pump by 200 μM vanadate abolished—as expected—any electric current transient upon ATP release from caged ATP

Pase, the two principal enzyme conformations could be identified (Helmich-de Jong et al. 1987), and the stoichiometry was found to be 2 H$^+$/2 K$^+$/1 ATP (Faller et al. 1982). The resulting overall electroneutrality leads to experimental difficulties in the performance of detailed studies of the ion transport by the H,K-ATPase. The fact that in each half-cycle of the pumping scheme two monovalent cations traverse the membrane has to result in a detectable electrogenic contribution when H$^+$ or K$^+$ transport are investigated separately. Fig. 9 shows a unambiguous proof that the ATP-induced H$^+$ transfer, H$_2$E$_1$ + ATP → ... → P-E$_2$ + 2 H$^+_{lum}$ + ADP, is accompanied by charge movement (van der Hijden et al. 1990). In these experiments, ATPase-containing vesicles prepared from pig stomach were capacitively coupled to a planar bilayer membrane, and enzyme phosphorylation was triggered by a flash-induced release of ATP from caged ATP. Corresponding experiments for the K$^+$ transporting branch of the pump cycle were not possible. Detailed time-resolved kinetical analyses beyond these data of van der Hijden and collaborators are still scarce (Stengelin et al. 1993).

In recent experiments, in which the styryl dye RH421 was applied to detect charge movements, the ion-binding sites were titrated with H$^+$ and K$^+$ ions in both principal conformations. In these experiments it could be shown that all binding and release steps are electrogenic, while enzyme phosphorylation by ATP produced no significant charge movement within the protein (Diller et al. 2003).

Energetic properties

Important for the understanding of the ion transport in ion pumps are, besides structural knowledge and kinetical properties, considerations on the energetics of transport. The knowledge of the "costs" in terms of free-enthalpy changes of the various reaction steps

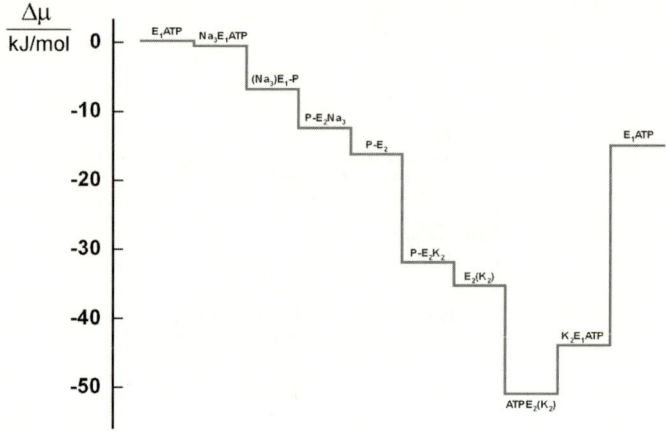

Fig. 10 Free energy levels of the individual states in the Na,K-ATPase pump cycle (Fig. 3). The energy levels were calculated according to Läuger (1991) on the basis of the kinetical parameters of Apell (1997). All levels refer to state $E_1 \cdot ATP$ as reference state

around the pump cycle provides at least clues to underlying processes which affect the protein structure and which facilitate the observed pump action.

Free energy levels

T.L. Hill showed that energy transduction in molecular machines like P-type ATPase is not the result of a single reaction step but of the cycle as a whole (Hill 1977, 1989). It is an interesting question, however, to what extent single reaction steps contribute to storage and consumption of the system's free energy. To gain access to such information, energy levels must be introduced for all the states of the pumping cycle which are long-lived states on the time scale of molecular motions and which are in equilibrium with respect to movements of the peptide backbone or amino acid side chains. Accordingly, the states can be treated as chemical species with a well-defined chemical potential which were introduced for those defined states of the ion pumps as "basic free energy levels" (Hill 1977). As has been shown earlier, the differences of free energy between two consecutive states can be determined from the forward and backward rate constants of the transition, or the corresponding equilibrium constant (Läuger 1991).

Analyses of free basic energy levels have been performed for ion pumps on a general level (Läuger 1984), for the SR Ca-ATPase on the basis of a less elaborate data base (Walz and Caplan 1988), and also for the Na,K-ATPase in great detail (Stein 1990; Apell 1997). Free basic energy calculations can be performed for all states around the Post–Albers cycle. Placing state E_1ATP (cf. Fig. 3) as initial level arbitrarily to zero, the sequence of states around the pumping cycle in the physiological mode led to a lower level after a cycle is completed (Fig. 10). For the sake of simplicity, the free energy gained by binding of ATP (in step $E_2(K_2) \rightarrow E_2(K_2)ATP$) is not implemented in this figure. If the energy difference between two successive states is near zero, the distribution between both states is close to its thermodynamic equilibrium. If the energy difference is negative, the reaction runs "downhill," i.e., it is dissipating energy; a positive energy difference indicates energy storage, either in the protein conformation or in the electrochemical potential difference between bulk phase and binding site(s) for the involved ion species.

Of special functional importance in the diagram shown in Fig. 10 is the free-energy decrease by enzyme phosphorylation and the subsequent conformational change, $Na_3E_1 \cdot ATP \rightarrow (Na_3)E_1\text{-}P \rightarrow P\text{-}E_2(Na_3)$, which effectively drains the states which bind Na^+ ions from the cytoplasm, thus producing dynamically the observed higher apparent affinity for Na^+ than for K^+ ions in state E_1, despite the values determined by steady-state titration experiments (Schneeberger and Apell 2001). Other significant "downhill" reaction steps are extracellular K^+ binding, $P\text{-}E_2 \rightarrow P\text{-}E_2K_2$, and ATP binding, $E_2(K_2) \rightarrow ATP \cdot E_2(K_2)$. In both cases the decrease of free energy is caused by ligand concentrations far above their affinities (which are ~0.1 mM for K^+ and 8 μM for ATP). A reduction of ligand concentrations would diminish the step size. This is true also in the most prominent "uphill" reaction, the K^+ release step in partial reaction $K_2E_1 \cdot ATP \rightarrow E_1 \cdot ATP$, in which the ions face a high cytoplasmic concentration of 150 mM from binding sites which have an affinity on the order of 0.1 mM. However, under turnover conditions K^+ release is not rate limiting because of the high rate constants of all ion release and binding steps in the reaction sequence $K_2E_1 \cdot ATP \rightarrow ... \rightarrow Na_3E_1 \cdot ATP$ so that it may be assumed to be in a dynamic equilibrium.

The effect of the actual ion concentrations on the energetics is extremely high in the case of the SR Ca-ATPase. In a contracted muscle, the ratio of the Ca^{2+} concentrations across the SR membrane is in the order of 50; in the relaxed muscle it is in the order of 30,000. When the free energy of transport is calculated under these conditions according to:

$$\Delta G_{transp} = \Delta \tilde{\mu}_{Ca} = 2RT \ln \frac{[Ca^+]_{lum}}{[Ca^+]_{cyt}} \tag{1}$$

one obtains ~20 kJ/mol (contracted state) and ~53 kJ/mol (relaxed state), assuming that there is no pH gradient across the membrane. These numbers have to be compared to the Gibbs Free Energy of ATP hydrolysis, $-\Delta G_{ATP}$, which was found to be ~63 kJ/mol in rat muscle (Meyer et al. 1982). The energy efficiency, $|\Delta G_{trans}/\Delta G_{ATP}|$, derived from these numbers increases from 32% to 84% when Ca^{2+} is stored back in the sarcoplasmatic reticulum. In the case of the Na,K-ATPase, the ion-concentration changes in the cytoplasm are not significant and the efficiency (at a membrane potential of −70 mV) is always around 70%.

However, the free-energy balance of the ATPases is not only affected by the ion concentrations, but also by their electrogenicity. Those partial reactions in the cycle which have been found to be electrogenic, i.e., especially ion binding and release steps, have rate constants (or equilibrium dissociation constants) which depend on the membrane potential and, as a consequence, their basic free energy levels can be modulated by the membrane potential (Läuger and Apell 1986; Apell 1997).

Temperature dependency

Another way to get information on the energetics of the ion pumps is from the temperature dependence of protein function. According to a description by Arrhenius, the activation energy for a chemical reaction, E_a, can be introduced by the empirical relation

Table 1 Energetics of selected reaction steps from rabbit kidney Na,K-ATPase. Enthalpy $H^\dagger$ and Entropy $S^\dagger$ were calculated from experimentally determined rate constants applying the theory of absolute reaction rates (unpublished data)

Rate limiting step of the reaction sequence	Enthalpy $H^\dagger$ in kJ/mol	Entropy $S^\dagger$ in J/mol K (at 25°C)
$Na_2E_1 \rightarrow Na_3E_1$	~20	−227
$Na_3E_1 \rightarrow Na_3E_1 \cdot ATP$	52	−226
$Na_3E_1 \cdot ATP \rightarrow (Na_3)E_1\text{-}P$	64	−224
$(Na_3)E_1\text{-}P \rightarrow P\text{-}E_2Na_3$	112	−219
$P\text{-}E_2K_2 \rightarrow E_2(K_2)$	65	−224
$E_2(K_2) \rightarrow E_2(K_2) \cdot ATP$	70	−224
$E_2(K_2) \cdot ATP \rightarrow K_2E_1 \cdot ATP$	66	−224

$$k = A \cdot \exp\left(-\frac{E_a}{RT}\right), \tag{2}$$

and determined from the slope of a so-called Arrhenius plot. Although we do not have a one-step reaction, but a cyclic reaction sequence in the ATPases, it is still possible, for example, to plot the hydrolyzing activity of the Na,K-ATPase in an Arrhenius plot (Apell 1997). However, the interpretation is not straightforward. There are good arguments to claim that the apparent activation energy is mainly that of the rate limiting step in the observed process. Therefore, it is possible that the Arrhenius plot may be bent when processes with different activation energy become rate limiting in the low and high temperature range and take over control of the scrutinized partial reaction (Apell 1997).

Instead of using the empirical concept of Arrhenius from the nineteenth century, which results in an activation energy representing approximately the height of the potential energy barrier of the rate-limiting reaction step, an alternative concept can be used. The theory of absolute reaction rates describes the rate constant of a reaction as a function of thermodynamically well-defined energies (Moore and Pearson 1981):

$$k = \frac{k_B T}{h} \cdot \exp\left(-\frac{G^\dagger}{RT}\right) = \frac{k_B T}{h} \cdot \exp\left(\frac{S^\dagger}{R}\right) \cdot \exp\left(-\frac{H^\dagger}{RT}\right) \tag{3}$$

This theory allows a determination of the contributions of entropy $S^\dagger$ and enthalpy $H^\dagger$ to a reaction step. With the assumption that the transition state is in equilibrium with the ground state, it is possible to draw the same conclusions from the magnitude of $S^\dagger$ and $H^\dagger$ on structural changes during a partial reaction as one can get from thermodynamic parameters of overall reactions.

The reaction steps of the Na,K-ATPase occur in an aqueous environment and, therefore, entropy changes are dominated by binding and release of water molecules or ions. If a negative entropy change is observed, it is caused mainly by immobilization of solvent molecules or ions in the protein. From so far unpublished experiments in our lab, $S^\dagger$ and $H^\dagger$ were determined for a number of partial reactions (Table 1). These are recent results and the project is still under development. However, it is obvious that only in one of the reaction steps which are accessible so far, a significantly higher enthalpy, $H^\dagger$, was determined: in the conformation transition from E_1 to E_2: $(Na_3)E_1\text{-}P \rightarrow P\text{-}E_2Na_3$. This step was also found to be the rate-limiting step in the Na^+ transport branch of the pump cycle. Remarkable also is the finding that the entropy changes in all steps analyzed are approximately the same, on the order of −225 J/molK, indicating that there exists no process in which more energy is "wasted" than in others.

The information which we hope to obtain finally from the thermodynamic properties of the pump process, namely from basic free energies, changes of entropy and enthalpy, will provide constraints for mechanistic models and support the development of microscopic descriptions of protein functions as well as of the coupling of enzymatic activity and transport.

Probing crucial amino acids

Before the structure of the SR Ca-ATPase became known, the most important tool to obtain information on structure–function relations was the investigation of pump proteins modified by mutagenesis. Numerous laboratories studied effects on enzymatic or transport properties of the ATPases which could be detected after an exchange of amino acids. By this technique, parts of the protein could be identified which are crucial for substrate binding, transport, or conformational changes. Such studies were performed with the SR Ca-ATPase in order to pin down amino acids involved in Ca^{2+} binding sites (Clarke et al. 1989a, Andersen 1995; Vilsen 1995) and these were eventually confirmed by the structure of the crystallized protein (Toyoshima et al. 2000). It was found that the two Ca^{2+} binding sites were formed by the side-chain oxygen atoms of Asn 768, Glu 771, Thr 799, Asp 800, and Glu 908 as site I and by Asn 796, Asp 800, Glu 309, and the main chain carbonyl oxygen atoms of Val 304, Ala 305, Ile 307 as site II. To couple ion-transport in/through the membrane domain and enzymatic activity in the N, P, and A domains, the cytoplasmic loop L67 between helix M6 and M7 also plays an important role.

In the case of other P-type ATPases, especially the Na,K-ATPase and H,K-ATPase, for which extensive functional studies are available, structural similarities to the Ca-ATPase may be exploited to identify elements specific for transport function (Sweadner and Donnet 2001). However, such comparisons should be used preferentially as a basis for conceptual considerations and to scout purposeful point mutations.

The era of mutagenesis as an investigative tool began after the sequences of ion pumps became available (Shull et al. 1985, 1988) and within a short period of time the number of contributions almost exploded. As expression systems for the mutated sequence, several assays are applied: cRNA injection into *Xenopus* oocytes is very useful for electrophysiological studies (Horisberger et al. 1991; Vasilets et al. 1991), while for biochemical studies overexpression of ATPases in yeast (Scheiner-Bobis and Farley 1994; Pedersen et al. 1996) or insect cell lines (Blanco et al. 1995; Gatto et al. 2001) is convenient. In such cell lines the use of a baculovirus infection system was very successful for the examination of the H,K-ATPase (Klaassen et al. 1993) and the Na,K-ATPase (Gatto et al. 2001).

A recent review on established structure–function relationships through site-directed mutagenesis (Jørgensen and Pedersen 2001) showed that in the transmembrane segments M4, M5, and M6 of the α subunit, at least nine amino acids could be identified which are important for binding of Na^+, K^+, or Tl^+. Most of them are homologous counterparts of the side chains which form the Ca^{2+} binding sites in the SR Ca-ATPase. In the yeast expression system with a renal $\alpha_1\beta_1\gamma$ enzyme, the crucial amino acids were: Glu 327, Asn 776, Glu 779, Asp 804, Thr 807, and Asp 808. Other amino acids could be shown to affect the dissociation constants for Mg^{2+} and ATP (Jørgensen and Pedersen 2001). The fifth and

sixth transmembrane domains were also recently investigated by cysteine-scanning mutagenesis, another powerful tool to identify the role of single amino acids (Guennoun and
Horisberger 2000, 2002).

Mutagenesis studies were also performed with the H,K-ATPase to determine amino acids in the transmembrane segments that are involved or affect ion binding and transport
(Hermsen et al. 2000, 2001; Asano et al. 2001; Rulli et al. 2001). In transmembrane domains M4, M5, and M6, four conserved, negatively charged amino-acid side chains were
found which are important for ion binding and thus for the pumping process: Glu 343, Glu
795, Glu 820, and Asp 824 (Sequence according to Shull and Lingrel 1986). It could be
demonstrated that a replacement of Glu 795 and Glu 820 by uncharged amino acid side
chains resulted in a phenotype with a constitutive ATPase activity in the absence of K^+
ions. It was suggested that the absence of the two negatively-charged side chains mimics
an occupation of the ion binding sides by K^+ ions, the state which would be the trigger for
enzyme dephosphorylation in the wild-type enzyme (Hermsen et al. 2001). In M6, an additional amino acid was identified that also affects K^+-dependent dephosphorylation, Leu
817 (Asano et al. 2001).

A demonstration of the close relationship of Na,K-ATPase and gastric H,K-ATPase is
made visible in chimera of both ion pumps. In a recent paper it was shown that substitu-
tion of three residues in the M4 segment of the Na,K-ATPase sequence with their H,K-
ATPase counterparts (Leu319Phe, Asn326Tyr, Thr340Ser) and replacing the M3-M4 loop
sequence with that of the H,K-ATPase α-subunit results in a protein that exhibits 50% of
its maximal ATPase activity in the absence of Na^+ ions when the assay is performed at
pH 6.0 (Mense et al. 2002). Twenty-one of the 29 amino acids that are thought to form
M4 are already identical in both ATPases.

A different approach to test amino acids and their close environment is the application
of specific oxidative cleavage mediated by Fenton chemistry with complexed transition
metal ions, such as Fe^{2+} complexed by ATP or Cu^{2+} by 4,7-diphenyl-1,10-phenantroline
(Goldshleger et al. 2001; Tal et al. 2001). With this method a region of interaction between the α and β subunits of the Na,K-ATPase could be identified, and short sequences
of four amino acids in transmembrane helices M1 and M3 that are in proximity to each
other near the cytoplasmic surface. Applying this technique to the H,K-ATPase led to results essentially identical to that for Na,K-ATPase (Shin et al. 2001).

Comparisons of transport mechanism
of SR Ca-ATPase, Na,K-ATPase and H,K-ATPase

Stimulated by the 3D structure of the SR Ca-ATPase, the close relation between P-type
ATPases is discussed to a great extent, especially for the three most prominent pumps, SR
Ca-ATPase, Na,K-ATPase, and H,K-ATPase. The most recent overview of this field will
be published in 2003 as a special volume of the *Annals of the New York Academy of Sci-
ences*, containing the *Proceedings of the 10th International Conference on Na,K-ATPase
and Related Cation Pumps, vol. 986 (2003)*.

Summarizing the above-discussed findings in this presentation, it can be proposed that
the high homology in amino-acid sequence and, most probably, a high similarity of the
quaternary structure of the three ATPases, is complemented by a single concept of the
transport mechanism which allows a description of all transport phenomena observed so

Fig. 11 Schematic representation of the ion transport pathway and ion binding sites in P-type AT-Pases. Access to the binding sites from the aqueous phases is controlled by gates that alternate in opening the pathway. The pump is in an occluded state when both gates lock the access channels. An open state of both gates would produce an ion-channel like behavior and is prohibited under physiological conditions

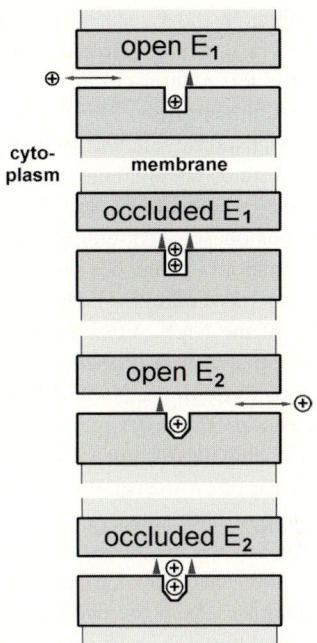

far. This concept is based on an alternate-access channel model discussed first by P. Läuger (Läuger 1979). The main features of the adapted mechanism are illustrated in Fig. 11, and they include (a) narrow access channels from both sides, (b) one gate per access channel, and (c) an ion binding moiety that is adapted specifically and differently in both principal conformations to the ions that are transported.

The transmembrane part of the pump, which is formed by ten helices of the α subunit and by the β subunit, is represented in Fig. 11 as a shaded box with a cavity inside the membrane. The cavity is connected to the bulk phase on both sides by narrow access channels. Each half channel possesses a gate that is controlled by the protein conformation (including changes induced by ATP binding and protein phosphorylation/dephosphorylation). A consequence of this arrangement is that ion binding and release is electrogenic on both sides. Such an electrogenicity was found for all three ion pumps, although in the case of the Na,K-ATPase the electrogenicity of cytoplasmic K^+ binding is hidden by transient H^+ binding (see the section entitled "Detailed ion transport mechanism"). With this finding the previously formulated constraint that the binding sites have to be twofold negatively charged is no longer strictly necessary. Nevertheless, negative (partial) charges in the binding sites are required to coordinate the dehydrated cations and to reduce the electric field of the cations in the largely nonpolar interior of the transmembrane part of the pump proteins.

The gates that block ion exchange between sites and aqueous phase are not necessarily mechanical barriers like floodgates. A look at the 3D structure of the crystallized Ca_2E_1 conformation of the Ca-ATPase reveals that there is indeed no steric obstacle in the pathway of the ions. Therefore, the gates which block the access channels have to be thought of as energy barriers produced by an arrangement of amino-acid side chains that do not allow dehydrated ions to be coordinated and thus make ion propagation energetically extremely unfavorable. In the E_1 states, the right-hand gate in the illustration in Fig. 11 is closed and access for ions is possible only from the cytoplasmic side. Correspondingly, in the states of E_2 the left gate is closed and ions can enter the binding sites only from the luminal (or extracellular) phase. The occluded states are transient states when the pumps run through their transport cycle and switch back and forth between both principal conformations. Occluded states are characterized by locking the binding moiety from both access channels. They are required to prevent ion-channel-like conformations that would short-circuit both aqueous phases. (Assuming that a pump transports typically 100 ions per second and an ion channel 10^7 ions per second, then a channel opening of about 10 μs would destroy the work of 1 s of pumping). However, under "unphysiological conditions" a channel-like behavior of the Na,K-ATPase could be produced by addition of Palytoxin, a potent marine toxin isolated from *Palythoa toxica*. In the presence of this compound, electric current fluctuations through the Na,K-ATPase were observed like those of a "classic" ion channel, with fluxes on the order of 10^6 ions per second (Artigas and Gadsby 2003). Because the channel behavior disappears when ATP is washed out, the channel-like behavior is not a consequence of an irreversible protein denaturation, but rather an induced and simultaneous opening of both gates.

The third feature of the mechanistic model is the binding moiety deep inside the membrane domain. In the case of the Ca-ATPase it was found to be inside about 30%–40% of the membrane thickness from the cytoplasmic surface (Toyoshima and Nomura 2002). In this moiety, ion-binding sites are formed in a way that they show differently high binding affinities for the transported ions in both principal conformations. In the case of the SR Ca-ATPase it can be seen that the low Ca^{2+} affinity $E_2(TG)$ state is produced by a severe disruption of the ion-coordination (cf. the section entitled "Structural properties"). A corresponding change may be proposed to occur by the conformation transition of the Na,K-ATPase, since an increase of the half-saturating Na^+ concentration was observed from ~7 mM (in E_1) to ~400 mM (in P-E_2; Heyse et al. 1994). After dissociation of the Na^+ ions, the moiety forming the binding-sites is thought to relax into a new equilibrium that is able to coordinate 2 K^+ ions with high affinity ($K_{1/2} \approx 0.1$ mM) (Bühler and Apell 1995).

When H^+ ions are bound to these pumps instead of hydronium ions (H_3O^+), the formation of a complex binding site with an up to sixfold coordination is not necessary; H^+ may bind to a single carboxylate anion of an amino-acid side chain. Therefore, "real" binding sites have be formed not necessarily for the P-$E_2 \rightarrow E_1$ branch of the SR Ca-ATPase and for the $E_1 \rightarrow$ P-E_2 branch of the H,K-ATPase. In Fig. 11 the different affinity for ions in the binding sites is indicated by the shape of the center cavity in the illustration.

Coupling of energetics and transport

An understanding of the transport mechanism of ion-translocating proteins, such as bacteriorhodopsin, the cytochrome-c oxidase, and the F_0F_1-ATPase was gained after functional observations and kinetic analyses could be correlated with structural information at atomic resolution. For example, studies on bacteriorhodopsin revealed how structural transitions of the chromophore, following absorption of a photon (energy uptake), induce a movement of the Schiff's base, which is used, in a precisely tailored surrounding, to transfer a proton across the central energy barrier in the transport pathway (Haupts et al. 1999; Lanyi and Luecke 2001). Recently, convincing proof was given that a mechanical movement of protein subunits, which form a rotor interacting with other parts of the F_0F_1-ATPase, are able to synthesize ATP by a "downhill" movement of protons through the F_0 unit or to pump protons "uphill" at the expense of ATP hydrolysis (Senior et al. 2002; Gaballo et al. 2002). In these cases, concepts of well-known macroscopic machines, like a mechanic pump or a water-driven mill, can be used, when scaled down, to describe for these highly specific machines dimensions of a few nanometers.

In the case of the P-type ATPases, such a comprehensive understanding is still lacking. Although structural details are now becoming available, as well as growing insight into functional properties on the basis of biophysical and biochemical studies as presented above, nevertheless, a convincing mechanistic concept of the energy transduction has not been formulated so far. However, ion pumps need not necessarily work as a scaled-down version of known macroscopic machines.

So far, a purely speculative concept can be based on the fact that both known conformations of the enzymatic part of the pump proteins, formed by the N, P, and A domains, switch between two shapes (Fig. 1), one similar to an open boxing-glove (E_1) and the other to a closed one (E_2) that surrounds (under physiological conditions) the Mg-P_i complex. In this "glove" the P domain represents the palm, the A domain the thumb, and the N domain the four fingers. And like the opening and closing of a fist requires movements of muscles and tendons in the forearm, the helices in the transmembrane part of the pump protein will move correspondingly with changes in the cytoplasmic domains. Driving forces for the propagation around the pumping cycle could be a series of substrate-protein interactions in which a binding/dissociation step produces transition states with an enhanced energy (by deformed electron orbitals) that relax by a conformational rearrangement into the next quasi-equilibrium state, ready for another substrate interaction. Such a mechanism is postulated, for example, in case of the Na,K-ATPase, where binding of the third Na^+ ion in E_1 acts as a trigger mechanism for enzyme phosphorylation (Schneeberger and Apell 1999).

However, the explicit energetics of P-type reveal, insofar as they have been determined up to now, that none of the known reaction steps constitute a "power stroke" which transfers the Gibbs free energy into the vectorial ion-moving process, analogous to that a mechanical pump. To reveal the "mechanical" representation (if there is any) of the process that shifts the pumps away from their thermodynamic equilibrium to keep them going is one of the challenges to be dealt with in the field of P-type ATPases in the years to come.

Acknowledgements This work was supported financially by the Deutsche Forschungsgemeinschaft (AP 45/4) and INTAS (Project 01–0224).

References

Abramson J, Svensson-Ek M, Byrne B, Iwata S (2001) Structure of cytochrome c oxidase: a comparison of the bacterial and mitochondrial enzymes. Biochim Biophys Acta 1544:1–9

Albers RW (1967) Biochemical aspects of active transport. Ann Rev Biochem 36:727–756

Altendorf K, Siebers A, Epstein W (1992) The KDP ATPase of Escherichia coli. Ann N Y Acad Sci 671:228–243

Andersen JP (1995) Dissection of the functional domains of the sarcoplasmic reticulum Ca^{2+}-ATPase by site-directed mutagenesis. Bioscience Reports 15:243–261

Andersen JP, Vilsen B (1988) Overview: subunit interaction and conformational states. Ca-ATPase and Na,K-ATPase compared. Prog Clin Biol Res 268A:603–622

Anner BM, Lane LK, Schwartz A, Pitts BJR (1977) A reconstituted $Na^+ + K^+$ pump in liposomes containing purified $(Na^+ + K^+)$-ATPase from kidney medulla. Biochim Biophys Acta 467:340–345

Anner BM, Moosmayer M (1981) Preparation of Na,K-ATPase-containing liposomes with predictable transport properties by a procedure relating the Na,K-transport capacity to the ATPase activity. J Biochem Biophys Meth 5:299–306

Apell H-J (1997) Kinetic and energetic aspects of Na⁺/K⁺-transport cycle steps. Ann N Y Acad Sci 834:221–230

Apell H-J, Bersch B (1987) Oxonol VI as an optical indicator for membrane potentials in lipid vesicles. Biochim Biophys Acta 903:480–494

Apell H-J, Diller A (2002) Do H⁺ ions obscure electrogenic Na⁺ and K⁺ binding in the E₁ state of the Na,K-ATPase? FEBS Lett 532:198–202

Apell H-J, Karlish SJD (2001) Functional properties of Na,K-ATPase, and their structural implications, as detected with biophysical techniques. J Membr Biol 180:1–9

Apell H-J, Läuger P (1986) Quantitative analysis of pump-mediated fluxes in reconstituted lipid vesicles. Biochim Biophys Acta 861:302–310

Apell H-J, Solioz M (1990) Electrogenic transport by the *Enterococcus hirae* ATPase. Biochim Biophys Acta 1017:221–228

Apell H-J, Borlinghaus R, Läuger P (1987) Fast charge translocations associated with partial reactions of the Na,K-pump: II. Microscopic analysis of transient currents. J Membr Biol 97:179–191

Apell H-J, Marcus MM, Anner BM, Oetliker H, Läuger P (1985) Optical study of active ion transport in lipid vesicles containing reconstituted Na,K-ATPase. J Membr Biol 85:49–63

Apell H-J, Häring V, Roudna M (1990) Na,K-ATPase in artificial lipid vesicles. Comparison of Na,K and Na-only pumping mode. Biochim Biophys Acta 1023:81–90

Apell H-J, Roudna M, Corrie JE, Trentham DR (1996) Kinetics of the phosphorylation of Na,K-ATPase by inorganic phosphate detected by a fluorescence method. Biochemistry 35:10922–10930

Aravind L, Galperin MY, Koonin EV (1998) The catalytic domain of the P-type ATPase has the haloacid dehalogenase fold. Trends Biochem Sci 23:127–129

Artigas P, Gadsby DC (2003) The Na⁺/K⁺ pump's intrinsic ligand-gated ion channel, unmasked by palytoxin. Proc Natl Acad Sci U S A 100:501–505

Asano S, Tabuchi Y, Takeguchi N (1989) Monoclonal antibody HK4001 completely inhibits K⁺-dependent ATP hydrolysis and H+ transport of hog gastric H⁺,K⁺-ATPase. J Biochem (Tokyo) 106:1074–1079

Asano S, Io T, Kimura T, Sakamoto S, Takeguchi N (2001) Alanine-scanning mutagenesis of the sixth transmembrane segment of gastric H⁺,K⁺-ATPase α-subunit. J. Biol. Chem. 276:31265–31273

Beauge LA, Glynn IM (1978) Commercial ATP containing traces of vanadate alters the response of (Na⁺ + K⁺) ATPase to external potassium. Nature 272:551–552

Beaugé LA, Glynn IM (1979) Occlusion of K ions in the unphosphorylated sodium pump. Nature 280:510–512

Beguin P, Crambert G, Monnet-Tschudi F, Uldry M, Horisberger JD, Garty H, Geering K (2002) FXYD7 is a brain-specific regulator of Na,K-ATPase α 1-β isozymes. EMBO J 21:3264–3273

Berrebi-Bertran I, Robert P, Camelin JC, Bril A, Souchet M (2001) The γ-subunit of (Na⁺,K⁺)-ATPase: a representative example of human single transmembrane protein with a key regulatory role. Cell Mol Biol 47:285–296

Bielen FV, Glitsch HG, Verdonck F (1991) Dependence of Na⁺ pump current on external monovalent cations and membrane potential in rabbit cardiac purkinje cells. J Physiol 442:169–189

Blanco G, Koster JC, Sánchez G, Mercer RW (1995) Kinetic properties of the α2β1 and α2β2 isozymes of the Na,K-ATPase. Biochem 34:319–325

Blostein R (1979) Side-specific effects of sodium on (Na,K)-ATPase. Studies with inside-out red cell membrane vesicles. J Biol Chem 254:6673–6677

Blostein R (1983) The influence of cytoplasmic sodium concentration on the stoichiometry of the sodium pump. J Biol Chem 258:12228–12232

Boldyrev A, Lopina O, Prokopjeva V, Stubbs C, Quinn PJ (1983) The modulation of Ca-ATPase activity and protein-lipid interactions in the sarcoplasmic reticulum by ATP. Biochem Int 6:297–305

Borlinghaus R, Apell H-J, Läuger P (1987) Fast charge translocations associated with partial reactions of the Na,K-pump: I. Current and voltage transients after photochemical release of ATP. J Membr Biol 97:161–178

Bühler R, Apell H-J (1995) Sequential potassium binding at the extracellular side of the Na,K-pump. J Membr Biol 145:165–173

Bühler R, Stürmer W, Apell H-J, Läuger P (1991) Charge translocation by the Na,K-pump: I. Kinetics of local field changes studied by time-resolved fluorescence measurements. J Membr Biol 121:141–161

Butscher C, Roudna M, Apell H-J (1999) Electrogenic partial reactions of the SR-Ca-ATPase investigated by a fluorescence method. J Membr Biol 168:169–181

Cantley LC, Josephson L, Warner R, Yanagisawa M, Lechene C, Guidotti G (1977) Vanadate is a potent (Na,K)-ATPase inhibitor found in ATP derived from muscle. J Biol Chem 252:7421–7423

Capaldi RA, Aggeler R (2002) Mechanism of the F_1F_0-type ATP synthase, a biological rotary motor. Trends Biochem Sci 27:154–160

Carafoli E (1992) The Ca^{2+} pump of the plasma membrane. J Biol Chem 267:2115–2118

Chiesi M, Inesi G (1980) Adenosine 5'-triphosphate dependent fluxes of manganese and hydrogen ions in sarcoplasmic reticulum vesicles. Biochem 19:2912–2918

Clarke DM, Loo TW, Inesi G, MacLennan DH (1989a) Location of high affinity Ca^{2+}-binding sites within the predicted transmembrane domain of the sarcoplasmic reticulum Ca^{2+}-ATPase. Nature 339:476–478

Clarke RJ, Apell H-J (1989) A stopped-flow kinetic study of the interaction of potential-sensitive oxonol dyes with lipid vesicles. Biophys Chem 34:225–237

Clarke RJ, Apell H-J, Läuger P (1989b) Pump current and Na^+/K^+ coupling ratio of Na^+/K^+-ATPase in reconstituted lipid vesicles. Biochim Biophys Acta 981:326–336

Clarke RJ, Kane DJ, Apell H-J, Roudna M, Bamberg E (1998) Kinetics of Na^+-dependent conformational changes of rabbit kidney Na^+,K^+-ATPase. Biophys J 75:1340–1353

Cornelius F (1989) Uncoupled Na^+-efflux on reconstituted shark Na,K-ATPase is electrogenic. Biochem Biophys Res Commun 160:801–807

Cornelius F (1991) Functional reconstitution of the sodium pump. Kinetics of exchange reactions performed by reconstituted Na/K-ATPase. Biochim Biophys Acta 1071:19–66

Cornelius F, Mahmmoud YA, Christensen HR (2001) Modulation of Na,K-ATPase by associated small transmembrane regulatory proteins and by lipids. J Bioenerg Biomembr 33:415–423

Cornelius F, Møller JV (1991) Electrogenic pump current of sarcoplasmic reticulum Ca^{2+}-ATPase reconstituted at high lipid/protein ratio. FEBS Lett 284:46–50

de Meis L (1985) Role of water in processes of energy transduction: Ca^{2+}-transport ATPase and inorganic pyrophosphatase. Biochem Soc Symp 50:97–125

de Weer P, Gadsby DC, Rakowski RF (1988) Overview: stoichiometry and voltage dependence of the Na/K pump. Prog Clin Biol Res 268A:421–434

de Weer P, Gadsby DC, Rakowski RF (2000) The Na/K-ATPase: a current-generating enzyme. In: Taniguchi K, Kaya S (eds) The Na/K pump and related ATPases. Elsevier Science B.V., Amsterdam, pp 27–34

de Weer P, Gadsby DC, Rakowski RF (2001) Voltage dependence of the apparent affinity for external Na^+ of the backward-running sodium pump. J Gen Physiol 117:315–328

Demchenko AP, Apell H-J, Stürmer W, Feddersen B (1993) Fluorescence spectroscopic studies on equilibrium dipole-relaxational dynamics of Na,K-ATPase. Biophys Chem 48:135–147

Der A, Keszthelyi L (2001) Charge motion during the photocycle of bacteriorhodopsin. Biochemistry (Mosc) 66:1234–1248

Diller A, Vagin O, Sachs G, Apell H-J (2003) Electrogenic partial reactions of the gastric H,K-ATPase. Biophys J 84:263a

Dimroth P (1987) Sodium ion transport decarboxylases and other aspects of sodium ion cycling in bacteria. Microbiol Rev 51:320–340

Dimroth P, Matthey U, Kaim G (2000) Critical evaluation of the one- versus the two-channel model for the operation of the ATP synthase's F_0 motor. Biochim Biophys Acta 1459:506–513

Domaszewicz W, Apell H-J (1999) Binding of the third Na^+ ion to the cytoplasmic side of the Na,K-ATPase is electrogenic. FEBS Lett 458:241–246

Eisenrauch A, Bamberg E (1990) Voltage-dependent pump currents of the sarcoplasmic reticulum Ca^{2+}-ATPase in planar lipid membranes. FEBS Lett 268:152–156

Eisenrauch A, Grell E, Bamberg E (1991) Voltage dependence of the Na,K-ATPase incorporated into planar lipid membranes. Soc Gen Physiol Ser 46:317–326

Epstein W, Walderhaug MO, Polarek JW, Hesse JE, Dorus E, Daniel JM (1990) The bacterial Kdp K^+-ATPase and its relation to other transport ATPases, such as the Na^+/K^+- and Ca^{2+}-ATPases in higher organisms. Philos Trans R Soc Lond B Biol Sci 326:479–486

Faller L, Jackson R, Malinowska D, Mukidjam E, Rabon E, Saccomani G, Sachs G, Smolka A (1982) Mechanistic aspects of gastric $(H^+ + K^+)$-ATPase. Ann N Y Acad Sci 402:146–163

Faller LD, Smolka A, Sachs G (1985) The gastric H,K-ATPase. In: Martonosi AN (ed) Enzymes of biological membranes, vol 3. Plenum, New York, pp 431–448

Faller LD, Diaz RA, Scheiner-Bobis G, Farley RA (1991) Temperature dependence of the rates of conformational changes reported by fluorescein 5'-isothiocyanate modification of H^+,K^+-, and Na^+,K^+- ATPases. Biochemistry 30:3503–3510

Farley RA, Faller LD (1985) The amino acid sequence of an active site peptide from the H,K-ATPase of gastric mucosa. J Biol Chem 260:3899–3901

Fedosova NU, Cornelius F, Klodos I (1995) Fluorescent styryl dyes as probes for Na,K-ATPase reaction mechanism: significance of the charge of the hydrophilic moiety of RH dyes. Biochemistry 34:16806–16814

Fendler K, Grell E, Haubs M, Bamberg E (1985) Pump currents generated by the purified Na$^+$K$^+$-ATPase from kidney on black lipid membranes. EMBO J 4:3079–3085

Fendler K, van der Hijden H, Nagel G, de Pont JJ, Bamberg E (1988) Pump currents generated by renal Na$^+$K$^+$-ATPase and gastric H$^+$K$^+$-ATPase on black lipid membranes. Prog Clin Biol Res 268A:501–510

Fendler K, Jaruschewski S, Hobbs A, Albers W, Froehlich JP (1993) Presteady-state charge translocation in NaK-ATPase from Eel electric organ. J Gen Physiol 102:631–666

Ferreira ST, Coelho-Sampaio T (1996) Intrinsic fluorescence as a probe of structure-function relationships in Ca^{2+}-transport ATPases. Bioscience Reports 16:87–106

Fluhler E, Burnham VG, Loew LM (1985) Spectra, membrane binding, and potentiometric responses of new charge shift probes. Biochemistry 24:5749–5755

Forbush B III (1984a) An apparatus for rapid kinetic analysis of isotopic efflux from membrane vesicles and of ligand dissociation from membrane proteins. Anal Biochem 140:495–505

Forbush B III (1984b) Na$^+$ movement in a single turnover of the Na pump. Proc Natl Acad Sci USA 81:5310–5314

Forbush B III (1987) Rapid release of ^{42}K and ^{86}Rb from an occluded state of the Na,K-pump in the presence of ATP or ADP. J Biol Chem 262:11104–11115

Forgac M (1999) Structure and properties of the clathrin-coated vesicle and yeast vacuolar V-ATPases. J Bioenerg Biomembr 31:57–65

Froehlich JP, Heller PF (1985) Transient-state kinetics of the ADP-insensitive phosphoenzyme in sarcoplasmic reticulum: implications for transient-state calcium translocation. Biochemistry 24:126–136

Gaballo A, Zanotti F, Papa S (2002) Structures and interactions of proteins involved in the coupling function of the protonmotive F$_o$F$_1$-ATP synthase. Curr Protein Pept Sci 3:451–460

Gadsby DC (1984) The Na/K pump of cardiac cells. Ann Rev Biophys Bioeng 13:373–398

Gadsby DC, Kimura J, Noma A (1985) Voltage dependence of Na/K pump current in isolated heart cells. Nature 315:63–65

Gadsby DC, Rakowski RF, de Weer P (1993) Extracellular access to the Na,K pump: pathway similar to ion channel. Science 260:100–103

Garty H, Lindzen M, Scanzano R, Aizman R, Fuzesi M, Goldshleger R, Farman N, Blostein R, Karlish SJ (2002) A functional interaction between CHIF and Na-K-ATPase: implication for regulation by FXYD proteins. Am J Physiol Renal Physiol 283:F607–F615

Gatto C, McLoud SM, Kaplan JH (2001) Heterologous expression of Na$^+$-K$^+$-ATPase in insect cells: intracellular distribution of pump subunits. Am J Physiol Cell Physiol 281:C982–C992

Geering K (2001) The functional role of β subunits in oligomeric P-type ATPases. J Bioenerg Biomembr 33:425–438

Geering K, Crambert G, Yu C, Korneenko TV, Pestov NB, Modyanov NN (2000) Intersubunit interactions in human X,K-ATPases: role of membrane domains M9 and M10 in the assembly process and association efficiency of human, nongastric H,K-ATPase α subunits (ATP1al1) with known β subunits. Biochemistry 39:12688–12698

Gerencser GA (1996) The chloride pump: a Cl$^-$-translocating P-type ATPase. Crit Rev Biochem Mol Biol 31:303–337

Glitsch HG (2001) Electrophysiology of the sodium-potassium-ATPase in cardiac cells. Physiol Rev 81:1791–1826

Glitsch HG, Tappe A (1995) Change of Na$^+$ pump current reversal potential in sheep cardiac Purkinje cells with varying free energy of ATP hydrolysis. J Physiol 484:605–616

Glynn IM (1985) The Na$^+$,K$^+$-transporting adenosine triphosphatase. In: Martonosi AN (ed) Membrane transport. Plenum Press, New York, pp 35–114

Glynn IM, Richards DE (1982) Occlusion of rubidium ions by the sodium-potassium pump: its implications for the mechanism of potassium transport. J Physiol 330:17–43

Goldshlegger R, Karlish SJ, Rephaeli A, Stein WD (1987) The effect of membrane potential on the mammalian sodium-potassium pump reconstituted into phospholipid vesicles. J Physiol (Lond) 387:331–355

Goldshleger R, Shahak Y, Karlish SJD (1990) Electrogenic and electroneutral transport modes of renal Na/K ATPase reconstituted into proteoliposomes. J Membr Biol 113:139–154

Goldshleger R, Patchornik G, Shimon MB, Tal DM, Post RL, Karlish SJ (2001) Structural organization and energy transduction mechanism of Na$^+$,K$^+$-ATPase studied with transition metal-catalyzed oxidative cleavage. J Bioenerg Biomembr 33:387–399

Green NM, MacLennan DH (2002) Calcium callisthenics. Nature 418:598–599

Guennoun S, Horisberger JD (2000) Structure of the 5th transmembrane segment of the Na,K-ATPase α subunit: a cysteine-scanning mutagenesis study. FEBS Lett 482:144–148

Guennoun S, Horisberger JD (2002) Cysteine-scanning mutagenesis study of the sixth transmembrane segment of the Na,K-ATPase α subunit. FEBS Lett 513:277–281

Hartung K, Grell E, Hasselbach W, Bamberg E (1987) Electrical pump currents generated by the Ca^{2+}-ATPase of sarcoplasmic reticulum vesicles adsorbed on black lipid membranes. Biochim Biophys Acta 900:209–220

Hartung K, Froehlich JP, Fendler K (1997) Time-resolved charge translocation by the Ca-ATPase from sarcoplasmic reticulum after an ATP concentration jump. Biophys J 72:2503–2514

Haupts U, Tittor J, Oesterhelt D (1999) Closing in on bacteriorhodopsin: progress in understanding the molecule. Annu Rev Biophys Biomol Struct 28:367–399

Hebert H, Xian R, Thomsen K, Maunsbach AB (2000) Structure of renal Na,K-ATPase as observed by cryo-EM of 2-D crystals. In: Taniguchi K, Kaya S (eds) Na/K-ATPase and related ATPases. Elsevier, Amsterdam, pp 43–48

Hegyvary C, Post RL (1971) Binding of adenosine triphosphate to sodium and potassium ion-stimulated adenosine triphosphatase. J Biol Chem 246:5234–5240

Helmich-de Jong ML, van Emst-de Vries SE, de Pont JJ (1987) Conformational states of $(K^+ + H^+)$-ATPase studied using tryptic digestion as a tool. Biochim Biophys Acta 905:358–370

Hermsen HP, Koenderink JB, Swarts HG, de Pont JJ (2000) The carbonyl group of glutamic acid-795 is essential for gastric H^+,K^+-ATPase activity. Biochemistry 39:1330–1337

Hermsen HP, Swarts HG, Wassink L, Koenderink JB, Willems PH, de Pont JJ (2001) Mimicking of K^+ activation by double mutation of glutamate 795 and glutamate 820 of gastric H^+,K^+-ATPase. Biochemistry 40:6527–6533

Heyse S, Wuddel I, Apell H-J, Stürmer W (1994) Partial reactions of the Na,K-ATPase: determination of rate constants. J Gen Physiol 104:197–240

Hilgemann DW (1994) Channel-like function of the Na,K pump probed at microsecond resolution in giant membrane patches. Science 263:1429–1432

Hill, T. L. (1977) Free energy transduction in biology. Academic Press, New York

Hill, T. L. (1989) Free energy transduction and biochemical cycle kinetics. Springer, New York

Holmgren M, Rakowski RF (1994) Presteady-state transient currents mediated by the Na/K pump in internally perfused Xenopus oocytes. Biophys J 66:912–922

Holmgren M, Wagg J, Bezanilla F, Rakowski RF, de Weer P, Gadsby DC (2000) Three distinct and sequential steps in the release of sodium ions by the Na^+/K^+-ATPase. Nature 403:898–901

Horisberger J-D, Giebisch G (1989) Na-K pump current in the *Amphiuma* collecting tubule. J Gen Physiol 94:493–510

Horisberger J-D, Jaunin P, Good PJ, Rossier BC, Geering K (1991) Coexpression of α_1 with putative β_3 subunits results in functional Na^+/K^+ pumps in *Xenopus* oocytes. Proc Natl Acad Sci USA 88:8397–8400

Humphrey PA, Lupfert C, Apell HJ, Cornelius F, Clarke RJ (2002) Mechanism of the rate-determining step of the Na^+,K^+-ATPase pump cycle. Biochemistry 41:9496–9507

Inesi G (1987) Sequential mechanism of calcium binding and translocation in sarcoplasmic reticulum adenosine triphosphatase. J Biol Chem 262:16338–16342

Inesi G, de Meis L (1989) Regulation of steady state filling in sarcoplasmic reticulum. Roles of back-inhibition, leakage, and slippage of the calcium pump. J Biol Chem 264:5929–5936

Inesi G, Sagara Y (1992) Thapsigargin, a high affinity and global inhibitor of intracellular Ca^{2+} transport ATPases. Arch Biochem Biophys 298:313–317

Inesi G, Cantilina T, Yu X, Nikic D, Sagara Y, Kirtley ME (1992) Long-range intramolecular linked functions in activation and inhibition of SERCA ATPases. Ann N Y Acad Sci 671:32–47

Jaunin P, Jaisser F, Beggah AT, Takeyasu K, Mangeat P, Rossier BC, Horisberger JD, Geering K (1993) Role of the transmembrane and extracytoplasmic domain of β subunits in subunit assembly, intracellular transport, and functional expression of Na,K-pumps. J Cell Biol 123:1751–1759

Jørgensen PL (1974) Isolation of (Na^++K^+)-ATPase. Meth Enzymol 32:277–290

Jørgensen PL, Pedersen PA (2001) Structure-function relationships of Na^+· K^+, ATP, or Mg^{2+} binding and energy transduction in Na,K-ATPase. Biochim Biophys Acta 1505:57–74

Jørgensen PL, Jorgensen JR, Pedersen PA (2001) Role of conserved TGDGVND-loop in Mg^{2+} binding, phosphorylation, and energy transfer in Na,K-ATPase. J Bioenerg Biomembr 33:367–377

Kapakos JG, Steinberg M (1982) Fluorescent labeling of (Na^++K^+)-ATPase by 5-iodoacetamidofluorescein. Biochim Biophys Acta 693:493–496

Kapakos JG, Steinberg M (1986) Ligand binding to (Na,K)-ATPase labeled with 5-iodoacetamidofluorescein. J Biol Chem 261:2064–2069

Karlish SJ (1988) Use of formycin nucleotides, intrinsic protein fluorescence, and fluorescein isothiocyanate-labeled enzymes for measurement of conformational states of Na+,K+-ATPase. Methods Enzymol 156:271–277

Karlish SJ, Glynn IM (1974) An uncoupled efflux of sodium ions from human red cells, probably associated with Na-dependent ATPase activity. Ann N Y Acad Sci 242:461–470

Karlish SJ, Pick U (1981) Sidedness of the effects of sodium and potassium ions on the conformational state of the sodium-potassium pump. J Physiol 312:505–529

Karlish SJ, Yates DW, Glynn IM (1978) Conformational transitions between Na+-bound and K+-bound forms of (Na+ + K+)-ATPase, studied with formycin nucleotides. Biochim Biophys Acta 525:252–264

Karlish SJ, Lieb WR, Stein WD (1982) Combined effects of ATP and phosphate on rubidium exchange mediated by Na-K-ATPase reconstituted into phospholipid vesicles. J Physiol 328:333–350

Karlish, SJD, Yates DW (1978) Tryptophan fluorescence of (Na+ +K+)-ATPase as a tool for study of the enzyme mechanism. Biochim Biophys Acta 527:115–130

Keeling DJ, Laing SM, Senn-Bilfinger J (1988) SCH 28080 is a lumenally acting, K+-site inhibitor of the gastric (H+ +K+)-ATPase. Biochem Pharmacol 37:2231–2236

Kirley TL, Wang T, Wallick ET, Lane LK (1985) Homology of ATP binding sites from Ca²⁺ and (Na,K)-ATPases: comparison of the amino acid sequences of fluorescein isothiocyanate labeled peptides. Biochem Biophys Res Commun 130:732–738

Klaassen CH, Van Uem TJ, De Moel MP, De Caluwe GL, Swarts HG, de Pont JJ (1993) Functional expression of gastric H,K-ATPase using the baculovirus expression system. FEBS Lett 329:277–282

Klodos I, Forbush B, III (1988) Rapid conformational changes of the Na/K pump revealed by a fluorescent dye RH-160. J Gen Physiol 92:46a

Kühlbrandt W, Auer M, Scarborough GA (1998) Structure of the P-type ATPases. Curr Opin Struct Biol 8:510–516

Lafaire AV, Schwarz W (1985) Voltage-dependent, ouabain-sensitive current in the membrane of oocytes of *xenopus laevis*. In: Glynn IM, Ellory JC (eds) The sodium pump. The Company of Biologists, Ltd., London,pp 223–225

Lancaster CR (2002) A P-type ion pump at work. Nat Struct Biol 9:643–645

Lanyi JK, Luecke H (2001) Bacteriorhodopsin. Curr Opin Struct Biol 11:415–419

Läuger P (1979) A channel mechanism for electrogenic ion pumps. Biochim Biophys Acta 552:143–161

Läuger P (1984) Thermodynamic and kinetic properties of electrogenic ion pumps. Biochim Biophys Acta 779:307–341

Läuger, P. (1991) Electrogenic ion pumps. Sinauer Assoc., Sunderland, MA

Läuger P, Apell H-J (1986) A microscopic model for the current-voltage behavior of the Na,K-pump. Eur Biophys J 13:309–321

Läuger P, Apell H-J (1988) Voltage dependence of partial reactions of the Na⁺/K⁺ pump: predictions from microscopic models. Biochim Biophys Acta 945:1–10

Lederer WJ, Nelson MT (1984) Sodium pump stoichiometry determined by simultaneous measurements of sodium efflux and membrane current in barnacle. J Physiol 348:665–677

Lee AG, East JM (2001) What the structure of a calcium pump tells us about its mechanism. Biochem J 356:665–683

Lin SH, Smirnova IN, Kasho VN, Faller LD (1997) Eosin, energy transfer, and RH421 report the same conformational change in sodium pump as fluorescein. Ann N Y Acad Sci 834:442–444

Linnertz H, Kost H, Obsil T, Kotyk A, Amler E, Schoner W (1998a) Erythrosin 5′-isothiocyanate labels Cys549 as part of the low-affinity ATP binding site of Na⁺/K⁺-ATPase. FEBS Lett 441:103–105

Linnertz H, Urbanova P, Obsil T, Herman P, Amler E, Schoner W (1998b) Molecular distance measurements reveal an $(\alpha\beta)_2$ dimeric structure of Na⁺/K⁺-ATPase. J Biol Chem 273:28813–28821

Loew LM, Scully S, Simpson L, Waggoner AS (1979) Evidence for a charge-shift electrochromic mechanism in a probe of membrane potential. Nature 281:497–499

Lytton J, Westlin M, Hanley MR (1991) Thapsigargin inhibits the sarcoplasmic or endoplasmic reticulum Ca-ATPase family of calcium pumps. J Biol Chem 266:17067–17071

MacLennan DH, Brandl CJ, Korczak B, Green NM (1985) Amino-acid sequence of a Ca²⁺ + Mg²⁺-dependent ATPase from rabbit muscle sarcoplasmic reticulum, deduced from its complementary DNA sequence. Nature 316:696–700

MacLennan DH, Rice WJ, Green NM (1997) The mechanism of Ca²⁺ transport by sarco(endo)plasmic reticulum Ca²⁺-ATPases. J Biol Chem 272:28815–28818

Madeira VM (1978) Proton gradient formation during transport of Ca2+ by sarcoplasmic reticulum. Arch Biochem Biophys 185:316–325

McDonough AA, Geering K, Farley RA (1990) The sodium pump needs its β subunit. FASEB J. 4:1598–1605

Mense M, Dunbar LA, Blostein R, Caplan MJ (2000) Residues of the fourth transmembrane segments of the Na,K-ATPase and the gastric H,K-ATPase contribute to cation selectivity. J. Biol. Chem. 275:1749–1756

Mense M, Rajendran V, Blostein R, Caplan MJ (2002) Extracellular domains, transmembrane segments, and intracellular domains interact to determine the cation selectivity of Na,K- and gastric H,K-ATPase. Biochemistry 41:9803–9812

Meyer RA, Kuchmerick MJ, Brown TR (1982) Application of ^{31}P-NMR spectroscopy to the study of striated muscle metabolism. Am J Physiol 242:C1–11

Michel H (1999) Cytochrome c oxidase: catalytic cycle and mechanisms of proton pumping—a discussion. Biochemistry 38:15129–15140

Mitchell P, Moyle J (1974) The mechanism of proton translocation in reversible proton-translocating adenosine triphosphatases. Biochem Soc Spec Publ 4:91–111

Møller JV, Juul B, le Maire M (1996) Structural organization, ion transport, and energy transduction of P-type ATPases. Biochim Biophys Acta 1286:1–51

Moore JW, Pearson RG (1981) Kinetics and mechanism, 3rd ed. Wiley, New York

Nagel G, Bashi E, Slayman CL (1991) Spectral tagging of reaction substates in a sodium pump analogue: The plasma membrane H$^+$-ATPase of *Neurospora*. In: Kaplan JH, de Weer P (eds) The sodium pump: recent developments. The Rockefeller University Press, New York, pp 493–498

Nakao M, Gadsby DC (1986) Voltage dependence of Na translocation by the Na/K pump. Nature 323:628–630

Nelson, N. (1995) Organellar proton-ATPases. Springer Verlag, New York

Nishie I, Anzai K, Yamamoto T, Kirino Y (1990) Measurement of steady-state Ca^{2+} pump current caused by purified Ca^{2+}-ATPase of sarcoplasmic reticulum incorporated into a planar bilayer lipid membrane. J Biol Chem 265:2488–2491

Orlowski S, Champeil P (1991) The two calcium ions initially bound to nonphosphorylated sarcoplasmic reticulum Ca^{2+}-ATPase can no longer be kinetically distinguished when they dissociate from phosphorylated ATPase toward the lumen. Biochemistry 30:11331–11342

Papa S, Zanotti F, Gaballo A (2000) The structural and functional connection between the catalytic and proton translocating sectors of the mitochondrial F1F0-ATP synthase. J Bioenerg Biomembr 32:401–411

Pavlov KV, Sokolov VS (2000) Electrogenic ion transport by Na$^+$,K$^+$-ATPase. Membr Cell Biol 13:745–788

Pedersen M, Roudna M, Beutner S, Birmes M, Reifers B, Martin H-D, Apell H-J (2001) Detection of charge movements in ion pumps by a family of styryl dyes. J Membr Biol 185:221–236

Pedersen PA, Rasmussen JH, Jørgensen PL (1996) Expression in high yield of pig $\alpha1\beta1$ Na,K-ATPase and inactive mutants D369 N and D807 N in *Saccharomyces cerevisiae*. J Biol Chem 271:214–222

Peinelt C, Apell H-J (2002) Kinetics of the Ca^{2+}, H$^+$ and Mg^{2+} interaction with the ion-binding sites of the SR-Ca-ATPase. Biophys J 82:170–181

Peinelt C, Apell H-J (2003) Time-resolved partial reactions of the SR Ca-ATPase investigated with a fluorescent styryl dye. Ann N Y Acad Sci (in press)

Peluffo RD, Berlin JR (1997) Electrogenic K$^+$ transport by the Na$^+$-K$^+$ pump in rat cardiac ventricular myocytes. J Physiol 501:33–40

Pintschovius J, Fendler K (1999) Charge translocation by the Na$^+$/K$^+$-ATPase investigated on solid supported membranes: rapid solution exchange with a new technique. Biophys J 76:814–826

Post RL, Hegyvary C, Kume S (1972) Activation by adenosine triphosphate in the phosphorylation kinetics of sodium and potassium ion transport adenosine triphosphatase. J Biol Chem 247:6530–6540

Pratap PR, Robinson JD (1993) Rapid kinetic analyses of the Na$^+$/K$^+$-ATPase distinguish among different criteria for conformational change. Biochim Biophys Acta 1151:89–98

Rakowski RF (1989) Simultaneous measurement of changes in current and tracer flux in voltage-clamped squid giant axon. Biophys J 55:663–671

Rakowski RF, Gadsby DC, de Weer P (1987) Voltage-clamp reversal of the sodium pump in dialyzed squid giant axons. Biol Bull 173:445–446

Rakowski RF, Vasilets LA, LaTona J, Schwarz W (1990) A negative slope in the current-voltage relationship of the Na$^+$/K$^+$ pump in *Xenopus Oocytes* produced by reduction of external [K$^+$]. J Membr Biol 121:171–187

Reinhardt R, Lindemann B, Anner BM (1984) Leakage-channel conductance of single (Na$^+$ + K$^+$)-ATPase molecules incorporated into planar bilayers by fusion of liposomes. Biochim Biophys Acta 774:147–150

Rephaeli A, Richards DE, Karlish SJ (1986) Electrical potential accelerates the E1P(Na)-E2P conformational transition of (Na,K)-ATPase in reconstituted vesicles. J Biol Chem 261:12437–12440

Rulli SJ, Louneva NM, Skripnikova EV, Rabon EC (2001) Site-directed mutagenesis of cation coordinating residues in the gastric H,K-ATPase. Arch Biochem Biophys 387:27–34

Sachs G, Besancon M, Shin JM, Mercier F, Munson K, Hersey S (1992) Structural aspects of the gastric H,K-ATPase. J Bioenerg Biomembr 24:301–308

Sachs JR (1977) Kinetic evaluation of the Na-K pump reaction mechanism. J Physiol 273:489–514

Saraste M, Sibbald PR, Wittinghofer A (1990) The P-loop—a common motif in ATP- and GTP-binding proteins. Trends Biochem Sci 15:430–434

Scarborough GA (1999) Structure and function of the P-type ATPases. Curr Opin Cell Biol 11:517–522

Schatzmann HJ (1953) Herzglykoside als Hemmstoffe für den aktiven Kalium und Natrium Transport durch die Erytrocytenmembran. Helv Physiol Pharmacol Acta 11:346–354

Schatzmann HJ (1989) The calcium pump of the surface membrane and of the sarcoplasmic reticulum. Annu Rev Physiol 51:473–485

Scheiner-Bobis G, Farley RA (1994) Subunit requirements for expression of functional sodium pumps in yeast cells. Biochim Biophys Acta 1193:226–234

Schneeberger A, Apell H-J (1999) Ion selectivity of the cytoplasmic binding sites of the Na,K-ATPase: I. Sodium binding is associated with a conformational rearrangement. J Membr Biol 168:221–228

Schneeberger A, Apell H-J (2001) Ion selectivity of the cytoplasmic binding sites of the Na,K-ATPase: II. Competition of various cations. J Membr Biol 179:263–273

Schwarz W Gu Q (1988) Characteristics of the Na$^+$/K$^+$-ATPase from *Torpedo californica* expressed in *Xenopus* oocytes: a combination of tracer flux measurements with electrophysiological measurements. Biochim Biophys Acta 945:167–174

Seidler NW, Jona I, Vegh M, Martonosi A (1989) Cyclopiazonic acid is a specific inhibitor of the Ca2+-ATPase of sarcoplasmic reticulum. J Biol Chem 264:17816–17823

Seifert K, Fendler K, Bamberg E (1993) Charge transport by ion translocating membrane proteins on solid supported membranes. Biophys J 64:384–391

Sen AK, Post RL (1964) Stoichiometry and localization of adenosine triphosphate-dependent sodium and potassium transport in the erythrocyte. J Biol Chem 239:345–352

Sen PC, Kapakos JG, Steinberg M (1981) Modification of (Na$^+$+K$^+$)-dependent ATPase by fluorescein isothiocyanate: evidence for the involvement of different amino groups at different pH values. Arch Biochem Biophys 211:652–661

Senior AE, Nadanaciva S, Weber J (2002) The molecular mechanism of ATP synthesis by F1F0-ATP synthase. Biochim Biophys Acta 1553:188–211

Shin JM, Goldshleger R, Munson KB, Sachs G, Karlish SJ (2001) Selective Fe^{2+}-catalyzed oxidative cleavage of gastric H$^+$,K$^+$-ATPase: implications for the energy transduction mechanism of P-type cation pumps. J Biol Chem 276:48440–48450

Shull GE, Greeb J (1988) Molecular cloning of two isoforms of the plasma membrane Ca^{2+}-transporting ATPase from rat brain. J Biol Chem 263:8646–8657

Shull GE, Lingrel JB (1986) Molecular cloning of the rat stomach (H$^+$ + K$^+$)-ATPase. J Biol Chem 261:16788–16791

Shull GE, Schwartz A, Lingrel JB (1985) Amino-acid sequence of the catalytic subunit of the (Na$^+$+K$^+$)ATPase deduced from a complementary DNA. Nature 316:691–695

Simons TJB (1974) Potassium: potassium exchange catalysed by the sodium pump in human red cells. J Physiol 237:123–155

Skou JC, Esmann M (1981) Eosin, a fluorescent probe of ATP binding to the (Na$^+$+K$^+$)-ATPase. Biochim Biophys Acta 647:232–240

Skou JC, Esmann M (1983) The effects of Na$^+$ and K$^+$ on the conformational transitions of (Na$^+$ + K$^+$)-ATPase. Biochim Biophys Acta 746:101–113

Slayman CL, Sanders D (1985) Steady-state kinetic analysis of an electroenzyme. Biochem Soc Symp 50:11–29

Sokolov VS, Apell H-J, Corrie JE, Trentham DR (1998) Fast transient currents in Na,K-ATPase induced by ATP concentration jumps from the P3-[1-(3′,5′-dimethoxyphenyl)-2-phenyl-2-oxo]ethyl ester of ATP. Biophys J 74:2285–2298

Stankiewicz PJ, Tracey AS, Crans DC (1995) Inhibition of phosphate-metabolizing enzymes by oxovanadium(V) complexes. Met Ions Biol Syst 31:287–324

Stein WD (1990) Energetics and the design principles of the Na/K-ATPase. J Theor Biol 147:145–159

Steinberg M, Karlish SJD (1989) Studies on conformational changes in Na,K-ATPase labeled with 5-iodoacetamidofluorescein. J Biol Chem 264:2726–2734

Stengelin M, Fendler K, Bamberg E (1993) Kinetics of transient pump currents generated by the (H,K)-ATPase after an ATP concentration jump. J Membr Biol 132:211–227

Stoeckenius W (1999) Bacterial rhodopsins: evolution of a mechanistic model for the ion pumps. Protein Sci 8:447–459

Stürmer W, Apell H-J, Wuddel I, Läuger P (1989) Conformational transitions and change translocation by the Na,K pump: comparison of optical and electrical transients elicited by ATP-concentration jumps J Membr Biol 110:67–86

Stürmer W, Bühler R, Apell H-J, Läuger P (1991) Charge translocation by the Na,K-pump: II. Ion binding and release at the extracellular face. J Membr Biol 121:163–176

Sweadner KJ, Donnet C (2001) Structural similarities of Na,K-ATPase and SERCA, the Ca^{2+}-ATPase of the sarcoplasmic reticulum. Biochem J 356:685–704

Sze H, Ward JM, Lai S (1992) Vacuolar H^+-translocating ATPases from plants: structure, function, and isoforms. J Bioenerg Biomembr 24:371–381

Tal DM, Capasso JM, Munson K, Karlish SJ (2001) Proximity of transmembrane segments M3 and M1 of the α subunit of Na^+,K^+-ATPase revealed by specific oxidative cleavage mediated by a complex of Cu^{2+} ions and 4,7-diphenyl-1,10-phenanthroline. Biochemistry 40:12505–12514

Therien AG Pu HX, Karlish SJ, Blostein R (2001) Molecular and functional studies of the γ subunit of the sodium pump. J Bioenerg Biomembr 33:407–414

Toyoshima C, Nomura H (2002) Structural changes in the calcium pump accompanying the dissociation of calcium. Nature 418:605–611

Toyoshima C, Nakasako M, Nomura H, Ogawa H (2000) Crystal structure of the calcium pump of sarcoplasmatic reticulum at 2.6 Å resolution. Nature 405:647–655

Tyson PA, Steinberg M, Wallick ET, Kirley TL (1989) Identification of the 5-iodoacetamidofluorescein reporter site on the Na,K-ATPase. J Biol Chem 264:726–734

Vagin O, Denevich S, Munson K, Sachs G (2002) SCH28080, a K^+-competitive inhibitor of the gastric H,K-ATPase, binds near the M5–6 luminal loop, preventing K^+ access to the ion binding domain. Biochemistry 41:12755–12762

van der Hijden HT, Grell E, de Pont JJ, Bamberg E (1990) Demonstration of the electrogenicity of proton translocation during the phosphorylation step in gastric H^+K^+-ATPase. J Membr Biol 114:245–256

Vasilets LA, Omay HS, Ohta T, Noguchi S, Kawamura M, Schwarz W (1991) Stimulation of the Na^+/K^+ pump by external $[K^+]$ is regulated by voltage-dependent gating. J Biol Chem 266:16285–16288

Vilsen B (1995) Structure–function relationships in the Ca^{2+}-ATPase of sarcoplasmic reticulum studied by use of the substrate analogue CrATP and site-directed mutagenesis. Comparison with the Na^+,K^+-ATPase. Acta Physiol Scand Suppl 624:1–146

Visser NV, van Hoek A, Visser AJ, Frank J, Apell H-J, Clarke RJ (1995) Time-resolved fluorescence investigations of the interaction of the voltage-sensitive probe RH421 with lipid membranes and proteins. Biochemistry 34:11777–11784

Walz D, Caplan SR (1988) Energy coupling and thermokinetic balancing in enzyme kinetics. Microscopic reversibility and detailed balance revisited. Cell Biophys 12:13–28

Wikstrom M (2000) Mechanism of proton translocation by cytochrome c oxidase: a new four-stroke histidine cycle. Biochim Biophys Acta 1458:188–198

Wuddel I, Apell H-J (1995) Electrogenicity of the sodium transport pathway in the Na,K-ATPase probed by charge-pulse experiments. Biophys J 69:909–921

Yu X, Carroll S, Rigaud J-L, Inesi G (1993) H^+ countertransport and electrogenicity of the sarcoplasmic reticulum Ca^{2+} pump in reconstituted proteoliposomes. Biophys J 64:1232–1342

Yu X, Hao L, Inesi G (1994) A pK change of Acidic residues contributes to cation countertransport in the Ca-ATPase of sarcoplasmic reticulum. J Biol Chem 269:16656–16661

Zhang P, Toyoshima C, Yonekura K, Green NM, Stokes DL (1998) Structure of the calcium pump from sarcoplasmic reticulum at 8-A resolution. Nature 392:835–839

Zhao J, Vasilets LA, Gu Q, Ishii T, Takeyas K, Schwarz W (1997) Transport activity of a chimeric Na^+,K^+-ATPase with Ca^{2+}/calmodulin binding domain from Ca^{2+}-ATPase in Xenopus oocytes. Ann N Y Acad Sci 834:372–375

Rev Physiol Biochem Pharmacol (2003) 150:36–90
DOI 10.1007/s10254-003-0017-x

H. Koepsell · B. M. Schmitt · V. Gorboulev

Organic cation transporters

Published online: 25 June 2003
© Springer-Verlag 2003

Abstract Over the last 15 years, a number of transporters that translocate organic cations were characterized functionally and also identified on the molecular level. Organic cations include endogenous compounds such as monoamine neurotransmitters, choline, and coenzymes, but also numerous drugs and xenobiotics. Some of the cloned organic cation transporters accept one main substrate or structurally similar compounds (oligospecific transporters), while others translocate a variety of structurally diverse organic cations (polyspecific transporters). This review provides a survey of cloned organic cation transporters and tentative models that illustrate how different types of organic cation transporters, expressed at specific subcellular sites in hepatocytes and renal proximal tubular cells, are assembled into an integrated functional framework. We briefly describe oligospecific Na^+- and Cl^--dependent monoamine neurotransmitter transporters (*SLC6*-family), high-affinity choline transporters (*SLC5*-family), and high-affinity thiamine transporters (*SLC19*-family), as well as polyspecific transporters that translocate some organic cations next to their preferred, noncationic substrates. The polyspecific cation transporters of the *SLC22* family including the subtypes OCT1-3 and OCTN1-2 are presented in detail, covering the current knowledge about distribution, substrate specificity, and recent data on their electrical properties and regulation. Moreover, we discuss artificial and spontaneous mutations of transporters of the *SLC22* family that provide novel insight as to the function of specific protein domains. Finally, we discuss the clinical potential of the increasing knowledge about polymorphisms and mutations in polyspecific organic cation transporters.

Introduction

"Organic cations" are organic molecules with a transiently or permanently positive net charge. This class includes many endogenous compounds of high physiological importance such as monoamine neurotransmitters, choline, and coenzymes, but also numerous

H. Koepsell (✉) · B. M. Schmitt · V. Gorboulev
Institut für Anatomie und Zellbiologie, Bayerischen Julius-Maximilians-Universität,
Koellikerstr. 6, 97070 Würzburg, Germany
e-mail: Hermann@Koepsell.de

drugs and xenobiotics. Because most organic cations are not membrane permeant, their intestinal absorption, tissue distribution, and biliary, renal, and intestinal excretion largely depends on membrane transporters. Moreover, several cation channels are permeable for small organic cations, but their contribution to these processes is not understood (Akaike et al. 1984; McCleskey and Almers 1985). Considering that endogenous organic cations can play various roles in different tissues, it is plausible that the evolution of multiple organic cation transporters served to allow for differential control in individual organs.

Transport of organic cations has been studied for more than forty years employing various approaches, including transport measurements in intact animals, isolated organs, tissue slices, perfused renal tubules, and isolated plasma membrane vesicles (for reviews, see Eisenhofer 2001; Elferink et al. 1995; Graefe and Bönisch 1988; Roch-Ramel et al. 1992; Turnheim and Lauterbach 1977; Ullrich 1994). The development of molecular biology techniques such as expression cloning allowed the identification and characterization of numerous organic cation transporters over the last fifteen years. The cloned transporters may be grouped into oligospecific and polyspecific transporters. For oligospecific transporters, there is a single main substrate, although compounds with closely related structures or smaller size may be also accepted. In contrast, polyspecific transporters translocate a number of organic cations of different molecular structures (Dresser et al. 2001; Koepsell 1998, 1999; Zhang et al. 1998a). Oligospecific cation transporters include the Na^+-cotransporters for neurotransmitters (Blakely et al. 1994; Giros et al. 1991, 1992; Kanner 1994; Kilty et al. 1991; Pacholczyk et al. 1991; Shimada et al. 1991), vesicular and plasma membrane cotransporters for choline (Apparsundaram et al. 2000, 2001; Okuda et al. 2000), and high-affinity transporters for thiamine (Rindi and Laforenza 2000). Polyspecific transporters that translocate organic cations may be further subdivided into transporters that preferentially translocate organic cations ("polyspecific organic cation transporters") and transporters that preferentially transport compounds other than organic cations but may translocate some organic cations, for example transporters of the SLC21 family that mainly translocate anions (van Montfoort et al. 1999).

This review provides a survey of all transporters that translocate organic cations. Among these, the Na^+- and Cl^--dependent monoamine neurotransmitter transporters, the choline transporters, and the high-affinity thiamine transporter are described briefly, whereas a more detailed discussion is dedicated to polyspecific transporters that predominantly translocate organic cations. The latter ones belong to a large transporter family named organic cation transporter (OCT)-family, or solute carrier family 22 (SLC22), which in turn is a member of the major facilitator superfamily (MFS; Burckhardt and Wolff 2000; Gründemann et al. 1998a; Koepsell 1998, 1999; Marger and Saier 1993; Pao et al. 1998; Schömig et al. 1998). We will point out the distributions, functional properties, and substrate- or inhibitor specificities, with particular emphasis on recent findings concerning the transport mechanism of the electrogenic cation transporters from this family because these have been characterized in some detail. We attempt to synthesize the current knowledge into a comprehensive synopsis of the physiological roles of organic cation transporters in liver and kidney. We also discuss the regulation of these transporters, and the structure–function studies using in vitro-mutagenesis. Finally, we report on the exciting recent findings concerning polymorphisms, mutations, and genetic diseases associated with the human organic cation transporters which underscore the biomedical importance of these transporters.

Oligospecific transporters that translocate organic cations

Monoamine neurotransmitter transporters

Specific Na⁺-driven cotransport systems in nerve terminals serve to retrieve monoamine neurotransmitters from the synaptic clefts and thus terminate neurotransmission. Specific Na^+-driven cotransporters exist for the monoamine neurotransmitters norepinephrine (NET), dopamine (DAT) and serotonin (SERT; Table 1; Bruss et al. 1993; Gelernter et al. 1993; Giros et al. 1992; Lesch et al. 1993; Ramamoorthy et al. 1993). These transporters are localized specifically at sympathetic, dopaminergic and serotoninergic synapses, respectively, and exhibit distinct substrate and inhibitor specificities (for reviews see Borowsky and Hoffman 1995; Eisenhofer 2001; Hoffman et al. 1998; Masson et al. 1999). The Na^+/monoamine neurotransmitter cotransporters belong to the SLC6-family of Na^+- and Cl^--dependent transporters; this family also comprises several amino acid transporters. By terminating the activation of the respective receptors in the postsynaptic membranes, Na^+/monoamine neurotransmitter cotransporters modulate neural transmission and are an integral component of intracellular metabolizing and recycling systems.

The expression of these transporters is not restricted to the brain. For example, the norepinephrine transporter NET is expressed in noradrenergic neurones, in chromaffin cells of the adrenal medulla, in the syncytiotrophoblast of the placenta, and in the capillary endothelial cells of the lung (Kippenberger et al. 1999; Phillips et al. 2001). In peripheral neurones and adrenal medulla, NET is responsible for the reuptake of norepinephrine and epinephrine and in the placenta, the NET helps to clear catecholamines from the fetal circulation (Bzoskie et al. 1995, 1997; Eisenhofer 2001; Nguyen et al. 1999). In the lung, NET is involved in the clearance of catecholamines which are metabolized by catechol-*O*-methyl-transferase (COMT) and monoamine oxidase (MAO; Bryan-Lluka et al. 1992; Catravas and Gillis 1980; Nicholas et al. 1974)

The gastrointestinal tract houses cells that secrete large amounts of dopamine. The dopamine transporter DAT is a downstream element of this signalling system (Eisenhofer et al. 1997; Goldstein et al. 1995). DAT is expressed in gastric parietal cells, in the endothelium of gastric mucosal blood vessels, in pancreatic duct cells, and in the endothelium of pancreatic venous blood vessels (Mezey et al. 1996, 1998). Receptor binding studies and results obtained with transgenic mice suggest that DAT is involved in the regulation of pancreatic bicarbonate secretion and gut motility (Mezey et al. 1999; Walker et al. 2000).

The serotonin transporter SERT is expressed not only in serotoninergic neurones in the central nervous system (CNS), but also in neurones of the enteric nervous system (ENS), and in intestinal mucosal cells (Chen et al. 1998; Hoffman et al. 1998; Walker et al. 1988). In the gut, SERT mediates the reuptake of serotonin that has been produced and released by enterochromaffine (EC) cells. These cells respond to luminal stimuli by releasing serotonin, which in turn activates adjacent primary afferent nerve fibers (Vialli 1966). Uptake of interstitial serotonin via SERT into neurones and intestinal mucosal cells terminates the activation of these serotoninergic neurones and prevents spill-over of serotonin into portal vein and liver. Defects in the monoamine neurotransmitter may lead to profound behavioral and neurochemical changes (Bengel et al. 1998; Gainetdinov et al. 1999).

Table 1 Oligospecific transporters from human that translocate organic cations

Name	HGNC symbol	Locus	Major substrate	K_m (μM)	Tissues	Accession no.	Reference
hNET1 (also NAT1)	SLC6A2	16q12.2	Norepinephrine (Na$^+$/Cl$^-$ dependent)	0.77	Central and peripheral noradrenergic neurones, adrenal medulla (chromaffine cells), placenta (syncytiotrophoblast), lung (capillary endothelial cells)	NM_001043	Pacholczyk et al. 1991; Bruss et al. 1993; Gelernter et al. 1993
hDAT	SLC6A3	5p15.3	Dopamine (Na$^+$/Cl$^-$ dependent)	2.4	Dopaminergic neurones (synapse), stomach (parietal cells, endothelium of mucosal vessels), pancreas (duct cells, endothelium of veins)	NM_001044, S44626	Giros et al. 1992
hSERT	SLC6A4	17q11.1-q12	Serotonin (Na$^+$/Cl$^-$ dependent)	0.46	Serotoninergic neurones, platelets, placenta, lung	NM_001045, X70697	Lesch et al. 1993; Ramamoorthy et al. 1993
hChT1 (high-affinity choline transporter)	SLC5A7	2q12	Choline (Na$^+$/Cl$^-$ dependent)	1.2	Brain: putamen, spinal chord, medulla nerve fibers, keratinocytes, hair follicle	AF276871, NM_021815	Apparsundaram et al. 2000
hCTL1 (choline transporter-like protein 1)		9q22/33	Choline	1.2	Rat: spinal chord, brain, colon	AJ245620	ÓRegan et al. 2000
hVAChT (vesicular acetylcholine transporter)	SLC18A3	10q11.2	Acetylcholine (exchange for H$^+$)	~900	Brain, synaptic vesicles of ChAT	NM_003055, U10554	Erickson et al. 1994
ThTr1, TC1 (thiamine transporter 1)	SLC19A2	1q23.2–3	Thiamine (exchange for H$^+$)	~2.6	Brain, kidney, liver, intestine, pancreas, cochlear hair cells, colonocytes	AF135488, AF153330	Diaz et al. 1999; Fleming et al. 2001; Fleming et al. 1999
ThTr2 (thiamine transporter 2)	SLC19A3	2q37	Thiamine		Placenta, kidney, liver	AF283317, AF271633, NM_025243	Rajgopal et al. 2001; Eudy et al. 2000

Choline and acetylcholine transporters

All cells depend on the membrane-impermeant organic cation choline (and hence on cho-line transporters) for the synthesis of certain membrane constituents. Moreover, choline is of particular importance for cells that synthesize and secrete the transmitter acetylcholine. Cholinergic neurones are found in the cerebral cortex, brainstem, spinal chord, and auton-omous ganglia. They participate in the control of attention, sleep/wake cycle, voluntary movements, and multiple autonomous functions via their projections to other central neu-rones, autonomous neurones, or muscle fibers, and via local secretion in a paracrine or en-docrine mode. Altered neuronal choline transport has been implicated in Alzheimer's dis-ease (Francis et al. 1999; Isacson et al. 2002), Parkinson's disease (Calabresi et al. 2000), Down's syndrome (Isacson et al. 2002), schizophrenia (Hyde and Crook 2001), brain isch-emic events, and aging (Lockman and Allen 2002).

Analogously to monoaminergic neurones, the cholinergic neurones carry Na^+-driven symporters in the plasma membrane, and H^+-driven exchangers in the secretory vesicles. The plasma membrane uptake systems are specific for choline rather than acetylcholine because the active transmitter, acetylcholine, is synthesized only in the cytoplasm, and it is thus the precursor compound choline that needs to be imported from the extracellular compartment. The accumulation of acetylcholine in the secretory vesicles of cholinergic neurones, on the other hand, is brought about by an H^+-driven exchanger that is specific for acetylcholine; in contrast to the polyspecific vesicular uptake systems for monoamine transmitters (VMAT1 and VMAT2), this H^+/acetylcholine exchanger (VAChT) appears to be genuinely monospecific.

A plasma membrane choline transporter was cloned recently and termed ChT1 (Appar-sundaram et al. 2000). Orthologs of ChT1 are known from several mammalian species in-cluding man, rat, and mouse (Apparsundaram et al. 2001; Haberberger et al. 2002; Ko-bayashi et al. 2002). This transporter contains 13 potential membrane-spanning, hydropho-bic α-helices and belongs to the solute carrier family SLC5, together with other prominent members such as the Na^+-D-glucose cotransporter SGLT1 (Hediger et al. 1987) or the Na^+-iodide cotransporter NIS (Dai et al. 1996). ChT1 is a Cl^-- and Na^+-dependent cotrans-porter that utilizes the electrochemical potential of Na^+ to fuel the uptake of choline. Upon heterologous expression, ChT1 from man, rat, and pig mediates uptake of choline with high affinity (apparent K_m ~1–3 μM). ChT1 is expressed mainly in cholinergic neurones, but also in keratinocytes (Haberberger et al. 2002; Kobayashi et al. 2002; Lips et al. 2002; Okuda et al. 2000; Okuda and Haga 2000). Within the cholinergic neurones, ChT1 is lo-calized at the presynaptic membrane where it mediates the reuptake of choline from the synaptic cleft stemming from enzymatic hydrolysis of acetylcholine.

Because choline uptake is the rate-limiting step for the synthesis of acetylcholine and subjected to short-term regulation by neuronal activity (Haga 1971; Haga and Noda 1973; Kuhar and Murrin 1978; Simon and Kuhar 1975; Yamamura and Snyder 1972), ChT1 is an important modulator of cholinergic transmission. Recently, a polymorphism in the hu-man ChT1 transporter was identified that causes an amino acid exchange in the third trans-membrane domain with an associated 40–50% decrease of the maximal velocity; this loss-of-function mutation may predispose individuals for cholinergic dysfunction (Okuda et al. 2002).

Inside the neurones, acetylcholine is resynthesized and subsequently packed away into synaptic vesicles by a vesicular acetylcholine transporter (Eiden 1998; Parsons et al. 1993). A human vesicular acetylcholine transporter, termed hVAChT, was cloned (Erickson et al. 1994) and characterized (Bravo and Parsons 2002; Varoqui and Erickson 1996). This transporter, assigned the gene name SLC18A3, transports acetylcholine with a K_m of ~900 µM. Transport of acetylcholine by hVAChT is dependent on the H^+ gradient established by V-type ATPase and sensitive to nanomolar concentrations of the structural analogon L-vesamicol.

The distribution of VAChT mRNA largely overlaps that of choline acetyltransferase (ChAT), the enzyme required for acetylcholine biosynthesis, in peripheral and central cholinergic neurones, e.g., in forebrain, basal ganglia, brainstem, spinal chord, and autonomic ganglia of the intestinal and genitourinary tract. However, the additional expression of VAChT in nonneuronal sites such as T and B lymphocytes and vascular endothelial cells is increasingly appreciated (Kirkpatrick et al. 2001). Interestingly, the VAChT gene localizes to chromosome 10q11.2, which is also the location of the ChAT gene. The entire sequence of the human VAChT cDNA is contained uninterrupted within the first intron of the ChAT gene locus, implying a unique mechanism for regulation of two functionally coordinated proteins.

Altered expression of VAChT has been reported in various disorders including Alzheimer's disease (Ikeda et al. 2000; Kuhl et al. 1999), Huntington's disease (Kuhl et al. 1999), amyotrophic lateral sclerosis (Nagao et al. 1998), progressive supranuclear palsy (Suzuki et al. 2002), and type-II diabetes (Spangeus and El-Salhy 2001), although the significance of this association is unclear.

A further family of choline transporter is constituted by the so-called choline transporter-like proteins (CTL). To date, little is known about the four human members of this family, but the findings in other species strongly suggest an important role in human physiology. Choline transporter-like proteins were originally identified by complementation cloning in yeast: growth of a choline transport-deficient yeast strain could be restored by heterologous expression of a particular cDNA clone from the electric lobes of the ray *Torpedo marmorata*, and that transformant exhibited saturable, hemicholinium-3 inhibitable, and Na^+-independent uptake of ^3H-choline (K_m ~1 μM, IC_{50} ~0.5 μM; O'Regan et al. 2000). Homology searches revealed orthologs in *Drosophila*, *C. elegans*, *Dictyostelium*, and *A. thaliana*. No orthologs were found in prokaryotes, whereas orthologs with particularly high amino acid identities of ~70% were found in rat and human. In situ hybridization in mouse tissues revealed wide expression of mCTL1 inside and outside the nervous system. On the one hand, mCTL1 is expressed in cholinergic neurones such as the motor neurones of the spinal chord and the facial nucleus, which is compatible with a role analogous to the plasma membrane choline transporters hChT1. Strong expression was also found in oligodendroglia throughout the central nervous system (CNS). Outside the CNS, high levels of hCTL1 mRNA were detected in colonic epithelial cells, and in lung.

Knowledge about human CTLs is limited to nucleotide sequences, obtained either from expressed sequence tags (EST) in the case of homologs hCTL1 (Unigene accession no. Hs.179902), hCTL2 (Unigene accession no. Hs.167515), and hCTL3 (GenBank accession no. AA329432), or from conceptual translations of genomic DNA in the MHCIII region in the case of hCTL4 (AAD21813). The amino acid identity is ~20% between hCTL1 and hOCT1, and ~16% between hCTL2 and hOCT1. The CTL transporters from all species including human share the structural feature of a large, probably glycosylated exofacial loop between transmembrane domains 1 and 2 with the members of the OCT family.

Interestingly, the hCTL1 gene is localized very closely upstream in a locus at 9q31.2 that is responsible for familial dysautonomia, a disorder that presents autonomic and motor symptoms attributable to defects in central and peripheral cholinergic transmission. It is tempting to speculate, extrapolating the functional properties and tissue distribution of mouse CTL1 to the human, that this genetic disorder might be caused by disturbance of the 5′ region of the hCTL1 gene.

Thiamine transporters

The organic cation thiamine (vitamin B_1) is required in animal cells as a precursor of thiamine pyrophosphate, a coenzyme of several enzymes of the intermediary metabolism (Rindi and Laforenza 2000). Knowledge about thiamine transport is derived mainly from functional studies in native tissues and cells (Rindi and Laforenza 2000); only recently, thiamine transporters could be cloned and studied in isolation (Diaz et al. 1999; Eudy et al. 2000; Fleming et al. 1999, 2001).

Free thiamine is a hydrophilic quaternary ammonium compound that is present at micromolar concentrations in the intestinal lumen. It is taken up across the luminal membrane of enterocytes by a high-affinity, electroneutral H^+/thiamine exchanger; a similar transporter is found in brush border membranes of the renal proximal tubule where it mediates the reuptake of filtered thiamine, and in red blood cells. This transporter is inhibited by various thiamine analogs, but does not interact with classical substrates of organic cation transporters such as choline or tetraethylammonium (TEA). In the cytoplasm, thiamine exists mainly in various phosphorylated forms. Basolateral exit of thiamine appears to be Na^+- and ATP-dependent, but is understood only poorly. Carrier-mediated transport of thiamine phosphate was reported from the blood–brain barrier.

The first cloned thiamine transporter was identified in 1999 in an effort to identify the gene responsible for the rare genetic disorder of "thiamine-responsive megaloblastic anemia" (TRMA; Dutta et al. 1999; Fleming et al. 1999). The transporter, likely to represent the luminal thiamine uptake mechanism, was termed ThTr1, whereas its gene (SLC19A2) was grouped into one family together with the "reduced folate transporter" RFC-1 (gene name SLC19A1) based on sequence similarity. RFC-1 has no functional overlap with ThTr1 inasmuch as it transports exclusively anionic substrates via anion exchange, but a close physiological relation beyond the sequence similarity is nonetheless given both by the fact that the main substrate folic acid is another vitamin of the B group, and by the recent finding that RFC-1 also transports thiamine monophosphate (TMP). The structural similarity between these transporters supports the notion that the evolution of these transporters occurred only partially along the dividing line between anionic and cationic substrates, and underscores the importance of other structural or biological factors in that process; such a view is also justifiable and potentially fruitful with respect to the structure–function relationships observed in transporters for organic anions (OATs), and OCTs (see the section entitled "SLC22 family").

ThTr1 contains 497 amino acids and 12 potential membrane-spanning α-helices. Northern blot analyses showed that human ThTr1 is expressed in skeletal muscle, placenta, heart, liver, and kidney. The functional characterization in HeLa cells showed that the human ThTr1 transporter transports thiamine with a high affinity (K_m ~2.6 µM), is stimulated by an outwardly directed H^+-gradient, and operates independently of Na^+. Choline, (TEA), 1-methyl-4-phenylpyridinium (MPP), and cimetidine do not interact with ThTr1 (Dutta et al. 1999). As mentioned above, mutations in the ThTr1 gene on chromosome 1 are responsible for the genetic disease TRMA, but it has remained unclear in what way the mutations affect the functional properties of the transporter, how altered function of the transporter can produce symptoms as divergent as megaloblastic anemia, diabetes, and deafness, and why this syndrome presents so profoundly differently from the sequels of dietary thiamine deficiency (Fleming et al. 2001). In principle, the specific damage caused by the transporter defect might be set mainly during embryonic development, compatible with the known relation between vitamin B-deficiency and birth defects, or the transporter defect might be restricted to particular cells or tissues. Alternatively or in addition, thiamine transport might be the result of a collaborative effort of several transporters, in which case one should postulate the existence of further thiamine transporters.

Indeed, a further thiamine transporter was identified recently when it became clear that a third member of the SLC19 family (Eudy et al. 2000) functions with high specificity and affinity as a transporter for thiamine rather than for folate (Rajgopal et al. 2001). This transporter, termed ThTr2, is highly expressed in placenta, liver, and kidney and maps to the human chromosome 2q37 (Oishi et al. 2001). It has a high amino acid identity to RFC-1 (41%), but an even higher one to ThTr1 (58%). In accordance with the closer relation to the thiamine transporter ThTr1, the ThTr2 protein mediated uptake of ^{3}H-thiamine in transfected HeLa cells, but did not transport folic acid or methotrexate. Similarly, ^{3}H-thiamine uptake by ThTr2 was strongly inhibited by unlabeled thiamine, whereas several organic cations were ineffective. The activity of ThTr2 increased with pH, compatible with a H^+/thiamine exchange mechanism as demonstrated for ThTr1.

In the DBA/2J mouse, the ThTr2 gene maps closely to a seizure susceptibility locus, making ThTr2 a candidate for that disorder in that strain. No similar association is known for humans, but elucidation of the mouse pathology might shed light on the currently unknown physiological roles of ThTr2 in humans (Fleming et al. 2001).

The SLC22 family

Common structural properties of the family members

In 1994, we identified the polyspecific organic cation transporter rOCT1 (official gene symbol assigned by the Human Genome Nomenclature Committee SLC22A1) from rat kidney by expression cloning (Gründemann et al. 1994). Subsequent expression cloning of the rat organic anion transporter OAT1 (SLC22A6) from rat and flounder (Sekine et al. 1997; Sweet et al. 1997; Wolff et al. 1997) and homology cloning of further family members revealed that rOCT1 was the first prototypical member of a large transporter family within the major facilitator superfamily (MFS; Marger and Saier 1993; Pao et al. 1998; Sekine et al. 2000). The MFS superfamily comprises uniporters, symporters, and antiporters from bacteria, lower eukaryotes, plants, and mammals including drug resistance proteins, sugar facilitators, facilitators of Krebs cycle intermediates, organophosphate-phosphate antiporters, and oligosaccharide-H^+ symporters in 18 families. Most MFS transporters, including the members of the SLC22-family, share a predicted membrane topology with 12 predicted α-helical transmembrane domains and several common motifs (Fig. 1). In our previous reviews, we named the new transporter family OCT because OCT1 was the first identified member (Koepsell 1998; Koepsell et al. 1999). This nomenclature was not generally accepted, probably because it does not express the fact that this family comprises not only organic cation transporters (OCTs, OCTNs), but also OATs and

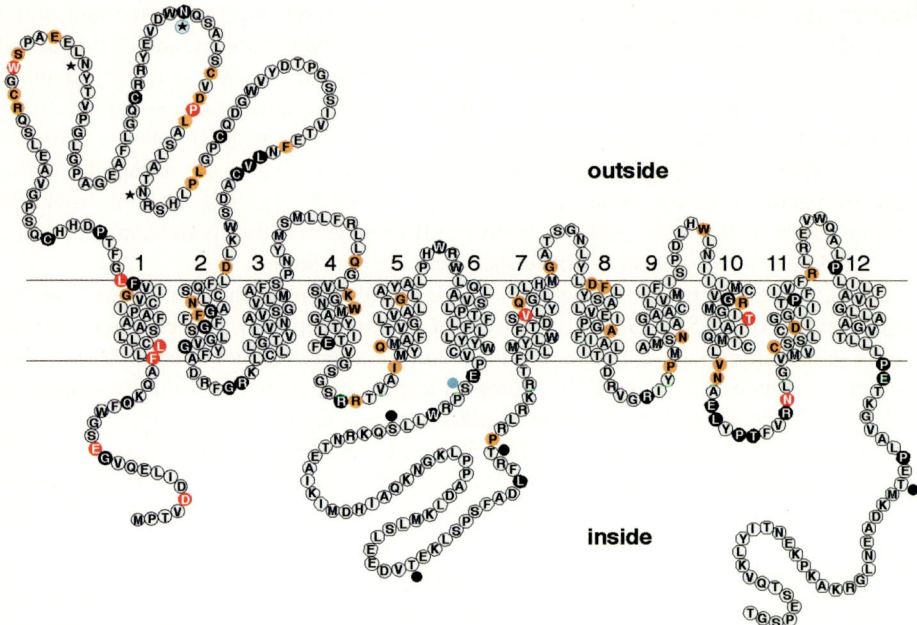

Fig. 1 The human electrogenic organic cation transporter hOCT1: amino acid sequence and current model of membrane topology. Amino acids (*a.a.*) that are conserved in particular subfamilies of the SLC22 transporter family are color-coded as follows: *black*, a.a. conserved in all members of the SLC22 family; *red*, a.a. conserved only in the mammalian organic cation transporters OCT1/OCT2/OCT3 and OCTN1/OCTN2; *orange*, a.a. conserved only in the electrogenic mammalian cation transporters OCT1/OCT2/OCT3; *blue*, consensus sequences for N-glycosylation or phosphorylation that are conserved in all functional members of the SLC22 family, or in all functional SLC22 family members except hCT2; *points*, consensus sequences for protein kinase C dependent phosphorylation sites in hOCT1; *asterisks*, N-glycosylation sites in hOCT1

transporters for the zwitterion carnitine (OCTN2, CT2) or for neutral compounds (OCTs, OATs). Schömig and collaborators advocated the name "amphiphilic solute facilitator" (ASF)-family (Gründemann et al. 1998a; Schömig et al. 1998), whereas other investigators used the term "organic ion transporter family" (Sekine et al. 2000) or grouped the various organic anion and cation transporters into separate families altogether (Enomoto et al. 2002a). However, the names "ASF" or "organic ion transporter family" merely replace the inconsistencies of the present terminology with new problems inasmuch many substrates either are not amphiphilic (such as tetramethylammonium), or not ionized at all (cimetidine at pH 8.5; Barendt and Wright 2002). A term that integrates OCTs, OATs, and OCTNs into one group and is not in logical conflict with the known properties of these transporters thus remains a desideratum in the field, given the generally acknowledged close relation between these transporters. Dropping the ambitious and possibly illusory claim to define a common property of all transported substrates in this family, we propose to designate all human OCTs, OATs and OCTNs as well as their respective orthologs from other species by the neutral term "SLC22," i.e., the official gene symbol assigned to these transporters by the "Human Genome Nomenclature Committee (HGNC)" (http://www.ge-ne.ucl.ac.uk/nomenclature/) on the basis of sequence homology.

Fig. 2 Phylogenetic tree of the human transporters that belong to the SLC22 family. Evolutionary relationships were calculated based on nucleotide sequence using the Felsenstein program Drawtree of the PHYLIP package (Felsenstein 1998). Distance along the branches is inversely related to the degree of sequence identity

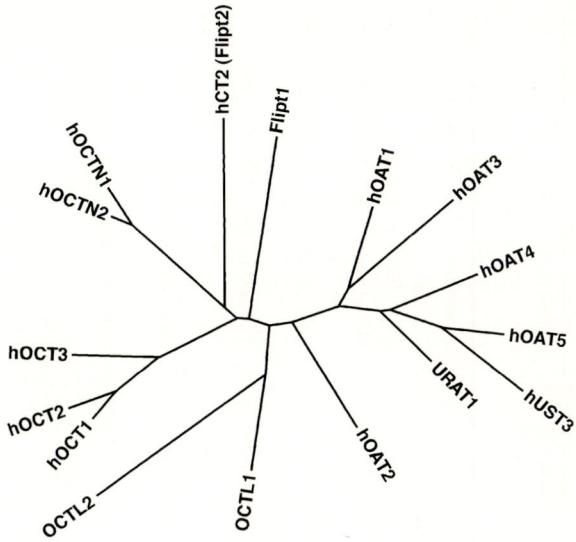

Members and predicted membrane topology

The known human members of the SLC22 family are presented in Table 2; Fig. 2 shows their phylogenetic relationship. Organic cations are transported by the three electrogenic organic cation transporter subtypes OCT1, OCT2, and OCT3 (Gorboulev et al. 1997; Gründemann et al. 1994, 1997, 1998a; Kekuda et al. 1998; Mooslehner and Allen 1999; Okuda et al. 1996; Schweifer and Barlow 1996; Terashita et al. 1998; Zhang et al. 1997a) and by the transporters OCTN1 and OCTN2 (Sekine et al. 1998; Tamai et al. 1997, 1998, 2000; Wu et al. 1998a, 2000a). A large group of transporters within the SLC22 family is engaged mainly in organic anion transport: Five organic anion transporters have been identified in humans, comprising OAT1–OAT4 and URAT1 (Cha et al. 2000, 2001; Enomoto et al. 2002a; Hosoyamada et al. 1999; Reid et al. 1998). OAT5, Fliptl, hUST3, OCTL1, and OCTL2 are gene products with unknown function (Eraly and Nigam 2002; Nishiwaki et al. 1998; Sun et al. 2001). With the exception of splice variants (Bahn et al. 2000; Urakami et al. 2002; Zhang et al. 1997b), all members of the SLC22 family contain 12 presumed transmembrane-spanning α-helices and two large hydrophilic loops. The first one connects the first and second transmembrane helix and could be assigned to the extracellular side in rOCT1 (Meyer-Wentrup et al. 1998), whereas the second one connects the sixth and seventh transmembrane helix and is probably localized intracellularly because it contains consensus sequences for protein kinase C-dependent phosphorylation (see proposed membrane topology of hOCT1 in Fig. 1). Of these phosphorylation sites, the one corresponding to serine 286 in rOCT1 (Koepsell et al. 1999) is conserved in almost all members of the SLC22 family.

Table 2 Transporters of the human SLC22A family

Name	a.a. Identity with hOCT1 (%)	Tissue expression	Gene locus	HUGO gene symbol	Nucleotide accession no.	References
hOCT1	100	Liver	6q26	SLC22A1	X98332, U77086	Zhang et al. 1997a; Koehler et al. 1997
hOCT2	70	Kidney, brain	6q26	SLC22A2	X98333, NM_003058	Koehler et al. 1997; Gorboulev et al. 1997; Busch et al. 1998
hOCT3	50	Liver, skeletal muscle, placenta, kidney, heart, lung, brain	6q26–27	SLC22A3	AJ001417, NM_021977	Gründemann et al. 1998a; Verhaagh et al. 1999; Wu et al. 2000b
hOCTN1	32	Kidney, skeletal muscle, placenta, prostate, heart	5q23.3	SLC22A4	AB007448, NM_003059	Tamai et al. 1997; Yabuuchi et al. 1999
hOCTN2	34	Kidney, skeletal muscle, placenta, heart, prostate, thyroid gland, pancreas, liver, etc.	5q23.3	SLC22A5	AF057164, NM_003060	Tamai et al. 1998; Wu et al. 1998a
hOAT1	32	Kidney, brain, placenta	11q12.3	SLC22A6	AF057039, NM_004790	Reid et al. 1998; Hosoyamada et al. 1999
hOAT2	30	Liver, kidney	6q21.1–2	SLC22A7	AF210455, AF097518, AY050498, NM_006672	Sun et al. 2001
hOAT3	31	Kidney, skeletal muscle, brain	11q12.3	SLC22A8	AF097491, NM_004254	Cha et al. 2001
hOAT4	30	Placenta, kidney	11q13.1	SLC22A11	AB026116, NM_018484	Cha et al. 2000
hOAT5	27	Liver	11q12.3	SLC22A10	AA705512	Sun et al. 2001
URAT1	32	Kidney	11q13.1	SLC22A12	AB071863	Enomoto et al. 2002a
Flipt1	29	Kidney, brain, skeletal muscle, heart, liver, placenta, lung, spleen	1p13.1		AY145501	Eraly et al. 2002
hCT2 (Flipt2)	30	Testis, liver, bone marrow, leukocytes, kidney	6q21–22.1		AB055798, AF268892, AY145502	Eraly et al. 2002; Gong et al. 2002; Enomoto et al. 2002b
hUST3	24	Liver	11q12.3		AA705512, AB062418	Sun et al. 2001
OCTL1	33	Kidney, small intestine, colon	3p22.2		AB010438	Nishikawa et al. 1998
OCTL2	28	Kidney, small intestine, colon	3p22.2		AB011082	Nishikawa et al. 1998

The electrogenic organic cation transporters OCT1, OCT2, and OCT3

Chromosomal localization and common functional properties of OCTs

The three subtypes of polyspecific electrogenic cation transporters, OCT1, OCT2, and OCT3, have been isolated from rat (Gründemann et al. 1994; Kekuda et al. 1998; Okuda et al. 1996), mouse (Mooslehner and Allen 1999; Schweifer and Barlow 1996; and Gen-Bank accession no. AF078750), and man (Gorboulev et al. 1997; Gründemann et al. 1998a; Zhang et al. 1997a). In addition, OCT1 was cloned from rabbit (Terashita et al. 1998), and OCT2 from rabbit and pig (Gründemann et al. 1997; Zhang et al. 2002). In human, the genes coding for OCT1, OCT2, and OCT3 are localized within a cluster on chromosome 6 (q26–27) (Gründemann and Schömig 2000; Gründemann et al. 1998a; Koehler et al. 1997; Eraly et al. 2002). Each of the three genes comprises 11 exons and 10 introns (Gründemann and Schömig 2000; Hayer et al. 1999; Verhaagh et al. 1999). On the protein level, individual transporter subtypes from human, mouse, and rat exhibit cross-species identities of 78–95% (OCT1), 81–91% (OCT2) and 87–93% (OCT3). Within a given species, the amino acid identities between different subtypes are 67–70% between OCT1 and OCT2, 47–57% between OCT1 and OCT3, and 49–51% between OCT2 and OCT3. Splice variants have been identified for OCT1 from rat and human (Hayer et al. 1999; Zhang et al. 1997b), and for human OCT2 (Urakami et al. 2002). One splice variant of rat OCT1, rOCT1A (Zhang et al. 1997b), lacks the first two transmembrane helices and the large extracellular loop that connects them. When expressed in *Xenopus* oocytes, rOCT1A exhibited TEA uptake with a Michaelis—Menten constant (K_m) value of 42 µM that was similar to the K_m of wild-type rOCT1 (95 µM, Gründemann et al. 1994), albeit at ~10% of the wild type's maximal transport rate. Four splice variants of human OCT1 showed no function (Hayer et al. 1999), whereas hOCT2A, a splice variant of human OCT2, has a truncated C-terminus lacking the last three proposed transmembrane domains, transported TEA with ~5% of the wild type's maximal rate, but revealed at higher affinity for a variety of cations (Urakami et al. 2002).

Within the human OCT transporters which translocate organic cations, 236 amino acids are conserved, whereas 413 amino acids are conserved within the human transporters OCTN1 and OCTN2 which translocate organic cations and are Na^+-cotransporters for carnitine. 107 amino acids are conserved between the human OCT transporters on the one hand and OCTN-transporters on the other. Considering transporters from all species, 34 amino acids are conserved between all members of the SLC22 family (see Fig. 1), they may have structural key functions. All OCT and OCTN transporters that are capable to transport organic cations (see Table 2) share 92 conserved amino acids. However, 82 of them are also present in some OATs of the SLC22 family, leaving only ten amino acids (see red amino acids in Fig. 1) which may convey cation-specific properties to organic ion transporters. The three OCT subtypes are related less closely among each other than the OCTN transporters. The higher degree of amino acid conservation of the OCTNs may be explained by the additional structural constraints imposed by the dual mode of operation of the OCTNs either as organic cation uniporter or as Na^+/carnitine cotransporter.

Thirty-six amino acids are conserved between OCT1–3 that are neither conserved in the OCTNs nor in the OATs (yellow in Fig. 1). All OCT and OCTN transporters contain several consensus sequences for protein kinase C- and protein kinase A-dependent phosphorylation. One of these sequences (Fig. 1, serine with blue point) is conserved throughout the whole SLC22 family excluding hCT1. The transporters of the SLC22 family furthermore contain consensus sequences for N-linked glycosylation within the large extra-cellular loop. One of these sequences is conserved in most family members, including the OCT and OCTN transporters (Fig. 1, blue star).

Cation transport by OCTs has been investigated in several heterologous expression systems including *Xenopus* oocytes (Busch et al. 1996a; Gründemann et al. 1994; Arndt et al. 2001; Budiman et al. 2000; Chen and Nelson 2000; Gorboulev et al. 1997; Kekuda et al. 1998; Nagel et al. 1997; Okuda et al. 1996, 1999; Terashita et al. 1998; van Montfoort et al. 2001; Wu et al. 2000b; Zhang et al. 1997a), human embryonic kidney (HEK) 293 cells (Arndt et al. 2001; Busch et al. 1996b; Gründemann et al. 1997, 1998a, 1998b; Mehrens et al. 2000; Pietig et al. 2001), HeLa cells (Wu et al. 1998b; Zhang et al. 1998b), MDCK cells (Sweet and Pritchard 1999a; Urakami et al. 1998, 2001), chinese hamster ovary (CHO-K1) cells (Barendt and Wright 2002), human retinal pigmented epithelium (HRPE) cells (Wu et al. 1998b), and others (Kimura et al. 2002; Pan et al. 1999; Takeda et al. 2002). Expressed transport was determined as uptake of radioactively labelled compounds, as electrical current across the entire oocyte membrane in the two electrode voltage-clamp configuration (for example see Arndt et al. 2001), or as current across giant patches excised from the oocyte plasma membrane in the patch clamp configuration (Budiman et al. 2000), and as cytoplasmic fluorescence change elicited by uptake of a fluorescent substrate (Mehrens et al. 2000).

The apparent K_m values for substrates and the apparent K_i values for inhibitors were largely independent from the employed expression system. In contrast, absolute transport rates differed largely between laboratories. Since the affinities for substrates and inhibitors are dependent on membrane potential and regulatory state of the transporter (Mehrens et al. 2000), different experimental conditions are likely to cause a certain variability of these parameters. In this respect, methods such as two-electrode voltage-clamp and patch-clamp have important advantages: because the membrane potential in these experiments is not only known, but also controlled efficiently (usually to −50 mV), this otherwise confounding source of variation can be eliminated such that the kinetic parameters obtained this way are more reproducible within one laboratory, and data from separate laboratories may be interpreted together.

Several properties are common to all OCTs and independent from subtype or species. (a) OCTs translocate a variety of organic cations with widely differing molecular structures, and are inhibited by other, nontransported compounds (Table 3). (b) OCTs translocate organic cations in an electrogenic manner. Electrogenicity of transport has been shown for the rat transporters rOCT1, rOCT2, and rOCT3 (Arndt et al. 2001; Busch et al. 1996a; Gründemann et al. 1994; Kekuda et al. 1998; Nagel et al. 1997; Okuda et al. 1999), and for the human transporters hOCT1 and hOCT2 (Busch et al. 1998; Dresser et al. 2000; Gorboulev et al. 1997). (c) OCTs operate independently from Na^+, Cl^-, and H^+ ions (Busch et al. 1996a; Gorboulev et al. 1997; Kekuda et al. 1998, and H. Koepsell et al., unpublished data). (d) OCTs are able to translocate cations across the plasma membrane in either direction. In addition to cation influx, cation efflux has been demonstrated for rOCT1, rOCT2, hOCT2, and rOCT3 (Busch et al. 1996a, 1998; Kekuda et al. 1998; Nagel et al. 1997).

Most substrates translocated by the OCT transporters are organic cations, but there are also several weak bases and noncharged compounds among the transported substrates (see Table 3 and Arndt et al. 2001; Dresser et al. 2001; Gründemann et al. 1999; Hayer-Zillgen et al. 2002; Urakami et al. 2001; Zhang et al. 1999, 2000). Transported substrates as well

Table 3 Substrates and inhibitors of human OCT and OCTN transporters. Data are derived from isotope uptake and uptake-inhibition experiments at physiological pH values using different expression systems (numbers in bold, K_m of transported substrates; numbers in parentheses, IC_{50} value for inhibition of transport of different substrate; charge symbols are: 0, uncharged and not ionized; +, permanent positive charge; –, permanent negative charge; +/–, zwitterion; (+), weak base; (–), weak acid)

Compound	Class	Charge	K_m or (IC_{50}) [µM]					References
			hOCT1	hOCT2	hOCT3	hOCTN1	hOCTN2	
Endogenous compounds								
Acetyl-L-carnitine	Metabolite	+					8.5	Ohashi et al. 1999
L-Carnitine	Metabolite	+/–					4.3 (4.8)	Tamai et al. 1998; Wagner et al. 2000
D-Carnitine	Metabolite	+/–					11 (98)	Ohashi et al. 1999; Wagner et al. 2000
Choline	Metabolite	+		210				Gorboulev et al. 1997
Corticosterone	Hormone	0	(7, 22)	(34)	(0.12, 0.29)			Zhang et al. 1998b; Hayer-Zillgen et al. 2002; Gründemann et al. 1998a
Dopamine	Neurotransmitter	(+)		390, 520				Busch et al. 1998
β-Estradiol	Hormone	0	(5.7)	(>30)	(2.9)			Hayer-Zillgen et al. 2002
Guanidine	Metabolite	+			(13,000) (6,201)			Gründemann et al. 1999; Wu et al. 2000b
Histamine	Biogenic amine	+		1,300	**180** (140)			Busch et al. 1998; Gründemann et al. 1999
Norepinephrine	Neurotransmitter	+		1,900	510			Gründemann et al. 1998a; Busch et al. 1998
Progesterone	Hormone	0	(3.1)	(27)	(4.3)			Hayer-Zillgen et al. 2002
Prostaglandin E$_2$	Hormone	–	0.66	0.03				Kimura et al. 2002
Prostaglandin F$_{2\alpha}$	Hormone	–	0.48	0.33				Kimura et al. 2002
Serotonin	Neurotransmitter	+		80				Busch et al. 1998
Drugs								
Acebutolol	β-Blocker	+	(96)					Zhang et al. 1998b
Acyclovir	Antiviral drug	(+)	151					Takeda et al. 2002
Amantadine	Anti-Parkinson	+		27 (23)				Busch et al. 1998
Cephaloridine	Antibiotic	+/–					230	Ganapathy et al. 2000
Cefepime	Antibiotic	+/–					1,700	Ganapathy et al. 2000

Table 3 (continued)

Compound	Class	Charge	K_m or (IC_{50}) [µM]					References
			hOCT1	hOCT2	hOCT3	hOCTN1	hOCTN2	
Cefoselis	Antibiotic	+/-					6,400	Ganapathy et al. 2000
Cimetidine	H2 rec. antagonist	(+)	(166)	8.6				Zhang et al. 1998b; Barendt et al. 2002
Clonidine	β2 Rec. antagonist	+	(0.55)		(373)			Zhang et al. 1998b; Wu et al. 2000b
Cocaine	Stimulant	+		(277)				H. Koepsell et al., unpublished data
Debrisoquine	Anti-hypertensive	+		7.3				H. Koepsell et al., unpublished data
Desipramine	Antidepressant	+	(5.4)	(16)	(14)			Zhang et al. 1998b; Gorboulev et al. 1997; Wu et al. 2000b
s(+)Disopyramide	Na+ ch. blocker	+	(30)					Zhang et al. 1998b
r(-)Disopyramide	Na+ ch. blocker	+	(15)					Zhang et al. 1998b
Emetine	Antiemetic	(+)					(4.2)	Wagner et al. 2000
Ganciclovir	Antiviral agent	(+)	516					Takeda et al. 2002
Imipramine	Antidepressant	+			(42)			Wu et al. 2000b
Indinavir	HIV prot. inhib.	(+)	(62)					Zhang et al. 2000
Memantine	Muscle relaxant	+		34				Busch et al. 1998
Mepiperphenidol	Anticholinergic	+		(4.8)				Gorboulev et al. 1997
Metformin	Antidiabetic	+		(2,010)	(1,700)			Dresser et al. 2002
Midazolam	Anesthetic	+	(3.7)					Zhang et al. 1998b
Nelfinavir	HIV prot. inhib.	(+)	(22)					Zhang et al. 2000
o-Methyl-isoprenaline	β Agonist	+	(>100)	(570)	(4.4)			Gorboulev et al. 1997; Hayer-Zillgen et al. 2002
Phenoxybenzamine	α Blocker	+	(2.7)	(4.9)	(6.1)			Hayer-Zillgen et al. 2002
Phenformin	Antidiabetic	+	(10)	(65)				Dresser et al. 2002
Prazosin	α Blocker	+	(1.8)	(>100)	(13)			Hayer-Zillgen et al. 2002
Procainamide	Na+ ch. blocker	+	(74), (107)	(50)	(738)			Gorboulev et al. 1997; Zhang et al. 1998b; Wu et al. 2000b; Zhang et al. 1999

Table 3 (continued)

Compound	Class	Charge	K_m or (IC_{50}) [µM]					References
			hOCT1	hOCT2	hOCT3	hOCTN1	hOCTN2	
Quinine	Antimalaria drug	(+)	(23)	(3.4)				Gorboulev et al. 1997; Zhang et al. 1998b
Quinidine	Na$^+$ ch. blocker	(+)	(18)					Zhang et al. 1998b
Ritonavir	HIV prot. inhib.	(+)	(5.2)					Zhang et al. 2000
Aquinavir	HIV prot. inhib.	(+)	(8.3)					Zhang et al. 2000
Vecuronium	Muscle relaxant	+	(127), (232), (237)					Zhang et al. 1998b; Zhang et al. 1999; Zhang et al. 1997a
Verapamil	Ca^{2+} ch. blocker	+	(2.9)	(206)				Zhang et al. 1998b; H. Koepsell et al., unpublished data
Xenobiotics and model cations								
Azidoprocainamide		+	101					Van Montfoort et al. 2001
Cyanine 863		+		(0.21)				Gorboulev et al. 1997
Decynium 22		+	(1.0–4.7)	(0.1), (1.1)	(0.09)			Gorboulev et al. 1997; Zhang et al. 1998b; Hayer-Zillgen et al. 2002; Zhang et al. 1997a
Disprocynium 24		+			(0.015)			Gründemann et al. 1998a
MPP (1-Methyl-4-phenylpyridinium)		+	15 (12)	19 (2.4)	47 (54)			Gorboulev et al. 1997; Zhang et al. 1998b; Wu et al. 2000b; Zhang et al. 1997a
Nicotine		+		(47)				H. Koepsell et al., unpublished data
N-methylquinine		+	20					Van Montfoort et al. 2001
N-methylquinidine		+	12					Van Montfoort et al. 2001
NMN (N-1-Methyl-nicotinamide)		+	(7,700)	340 (270)				Gorboulev et al. 1997; Zhang et al. 1998b
SKF550 (9-(Fluorenyl)-N-methyl-β-chloroethyl-amine)		+	(>0.3)	(0.1)	(0.05)			Hayer-Zillgen et al. 2002
Tetrabutylammonium		+	(30)	(52), (120)				Dresser et al. 2002; Zhang et al. 1999

Table 3 (continued)

Compound	Class	Charge	K_m or (IC$_{50}$) [μM]					References
			hOCT1	hOCT2	hOCT3	hOCTN1	hOCTN2	
Tetraethylammonium		+	229 (158–260)	76 (156)	(1,372)	436, 193		Gorboulev et al. 1997; Zhang et al. 1998b; Wu et al. 2000b; Dresser et al. 2002; Zhang et al. 1999; Tamai et al. 1997; Yabuuchi et al. 1999
Tetrahexylammonium		+	(3.0)					Zhang et al. 1999
Tetramethylammonium		+	(12,400)	(180), (150)				Gorboulev et al. 1997; Dresser et al. 2002; Dresser et al. 2000
Tetrapentylammonium		+	(7.5), (8.6)	(1.5)				Gorboulev et al. 1997; Zhang et al. 1998b; Zhang et al. 1999
Tetrapropylammonium		+	(90), (102)	(128)				Dresser et al. 2002; Zhang et al. 1999
Tributylmethylammonium		+	53 (66)					Zhang et al. 1999; Van Montfoort et al. 2001

as inhibitors of the OCT transporters may be endogenous compounds, drugs or xenobiotics. Transported endogenous substrates of the OCTs include the monoamine neurotransmitters acetylcholine, dopamine, serotonin, histamine, choline, and compounds such as
creatinine, guanidine, and thiamine. The steroid hormones corticosterone, deoxycorticos-
terone, and β-estradiol inhibit OCTs but are probably not transported themselves (H.
Koepsell et al., unpublished data). Many drugs and xenobiotics interact with the OCTs, either as transported substrates or as inhibitors. When the interaction of a compound with
OCT was tested indirectly via its effect on TEA or MPP uptake, it is important to keep in
mind that inhibition does not provide any clues as to whether the compound is transported
or not. Competitive binding of a second ligand can provide a sufficient explanation for an
observed inhibition. However, whether that second ligand is transported or not has no predictable effect on the uptake of the first ligand.

Examples for drugs that are transported by human OCTs include the histamine receptor
antagonist cimetidine (Barendt and Wright 2002; Zhang et al. 1998b), the antidiabetics
metformin and phenformin (Dresser et al. 2002), the antiviral agents acyclovir and ganciclovir (Takeda et al. 2002), the muscle relaxant memantine (Busch et al. 1998), and the
antiarrhythmic quinidine (van Montfoort et al. 2001). Many other drugs are known to inhibit human OCTs but have not been tested for transport (Table 3), including agonists and
antagonists of α-adrenoreceptors, β-adrenoreceptor antagonists, blockers of Na^+- and
Ca^{++}-channels, and antidepressants. Some cations like the xenobiotic MPP are transported
by all OCTs. Although organic cations are clearly the preferred ligands of the OCTs, several uncharged or anionic compounds are known to be inhibitors or even transported substrates of these transporters. For example, transport of the weak base cimetidine by hOCT2
is only partially dependent on the degree of ionization (Barendt and Wright 2002), and the
anionic prostaglandins are transported by hOCT1 (Kimura et al. 2002). Moreover, a number of anionic anti-inflammatory drugs such as indomethacin, diclofenac, ketoprofen, mefenamic acid, piroxicam, and sulindac are inhibitors of human OCT1 and OCT2 (Khamdang et al. 2002), and the organic anions probenecid, PAH, and α-ketoglutarate are inhibitors of the rat organic cation transporters rOCT1 and rOCT2 (Arndt et al. 2001).

Tissue distribution of OCTs

The tissue distribution of the OCT subtypes has been studied in some detail, but substantial gaps remain to be filled in order to fully understand the physiological roles of these
transporters. Expression of all cloned OCT transporters was studied in selected tissues by
Northern blotting (Gorboulev et al. 1997; Gründemann et al. 1994; Kekuda et al. 1998;
Okuda et al. 1996; Terashita et al. 1998; Wu et al. 2000b). For the rat OCTs, a more comprehensive and detailed analysis was carried out employing quantitative polymerase chain
reactions with reversely transcribed mRNAs from multiple tissues (Slitt et al. 2002). The
patterns of mRNA tissue distribution were found to be subtype-dependent within a given
species (Busch et al. 1998; Gorboulev et al. 1997; Gründemann et al. 1994; Kekuda et al.
1998; Okuda et al. 1996; Zhang et al. 1997a), and species-dependent for a given subtype
(Gorboulev et al. 1997; Gründemann et al. 1994).

In the species tested so far, OCT1 is mainly expressed in liver. In rodents, but not in
humans, OCT1 is also expressed strongly in the kidney (Gorboulev et al. 1997; Gründemann et al. 1994). In rats, high concentrations of OCT1mRNA were furthermore detected
in small intestine, skin and spleen (Slitt et al. 2002), whereas smaller concentrations of

OCT1 mRNA were found in many other tissues, for example in the brain and the mammary glands (Gerk et al. 2001).

OCT2 has a more restricted expression pattern than OCT1, being confined mainly to the kidney in human and other species (Gorboulev et al. 1997; Okuda et al. 1996). However, OCT2 was also detectable in human placenta and CNS neurones (Busch et al. 1998; Gorboulev et al. 1997) as well as in rat thymus (Slitt et al. 2002) and choroid plexus (Sweet et al. 2001).

The tissue expression pattern of OCT3 mRNA is relatively broad, with some species-specific particularities. In the human, OCT3 is mainly expressed in skeletal muscle, liver, placenta, kidney, and heart, and to a lesser extent in brain (Gründemann et al. 1998a; Wu et al. 2000b). In rats, OCT3 mRNA was mainly detected in placenta, small intestine, heart, and brain, and to a lesser extent in kidney, thymus, blood vessels, and skin (Kekuda et al. 1998; Slitt et al. 2002). In mice, OCT3 mRNA was found in placenta (Verhaagh et al. 1999). In situ-hybridization showed the presence of OCT3 mRNA in hippocampal and cerebellar neurones of rat and mice (Schmitt et al. 2003; Wu et al. 1998b); similarly, neuronal localization of OCT3 was also observed by single cell RT-PCR on neurones of the superior cervical ganglion of the rat (Kristufek et al. 2002).

Although these in situ hybridization experiments do not rule out an additional glial localization of OCT3 that is suggested by measurements of uptake$_2$-type transport in a glial cell line (Russ et al. 1996; Streich et al. 1996), the known neuronal localization is sufficient evidence to put into question the notion that OCT3 corresponds to a mainly extraneuronal norepinephrine transporter termed uptake$_2$ system. The uptake$_2$ system was originally described as a low-affinity, high-capacity uptake system for norepinephrine of cardiac myocytes, but was subsequently also found in vascular and nonvascular smooth muscle cells and kidney carcinoma cell lines (Iversen 1965; Schömig and Schönfeld 1990; Trendelenburg 1988). The uptake$_2$ system exhibited a distinct pattern of sensitivity to various inhibitors, including a very high sensitivity to corticosterone. Cation uptake by rat and human OCT3 showed similar sensitivities to some inhibitors as norepinephrine uptake into cardiac myocytes (Grohmann and Trendelenburg 1984; Iversen and Salt 1970) and human Caki 1-cells (Gründemann et al. 1998a; Schömig and Schönfeld 1990; Wu et al. 1998b). However, low-affinity transport of norepinephrine may be due to OCT1 and OCT2 as well (Busch et al. 1996b, 1998; Gründemann et al. 1998b); therefore, norepinephrine uptake cannot be assigned to OCT3 in tissues that express multiple OCT subtypes. In conclusion, the wide distribution of OCT3 including neurones, the long list of OCT3 substrates other than norepinephrine (Wu et al. 2000b), and the fact that norepinephrine is also a good substrate for other transporters such as OCT1 and OCT2 provide good reasons to abandon the simplistic and misleading habit of equating this transporter with "the" extraneuronal norepinephrine transporter. The original description of the extraneuronal uptake$_2$ system was based solely on functional criteria, but these might be satisfied by more than a single cloned transporter, both in theory and according to all available evidence.

Only few reports on the subcellular localization of OCT1–3 exist. In rat kidneys, rOCT1 and rOCT2 were localized at the basolateral membranes of renal proximal tubules (Karbach et al. 2000; Sugawara-Yokoo et al. 2000). Both transporter subtypes showed an overlapping distribution: rOCT1 was found in the S1 and S2 segments, whereas rOCT2 was localized to the S2 and S3 segments. Recently, a similar basolateral localization was shown for human OCT2 in the renal proximal tubules (Motohashi et al. 2002). In contrast to rat, however, human OCT2 is expressed in all three segments of the proximal tubules.

In rat liver, we localized rOCT1 protein to the sinusoidal membrane of hepatocytes, mainly those around the central veins (Meyer-Wentrup et al. 1998). Although no data are available on the immunolocalization of human OCT1 in liver, human OCT1 is likely to be expressed similarly at the sinusoidal membranes of the hepatocytes.

By immunohistochemistry, we detected OCT1 at the basolateral membrane of the enterocytes in the small intestine of rats (H. Koepsell et al., unpublished data), but also in serotoninergic neurones of the submucosal and myentric plexus in mice (Chen et al. 2001). In the small intestine, OCT1 is supposed to participate in the absorption and secretion of organic cations. In addition, OCT1 may mediate the reuptake of serotonin that has been released by EC cells (Chen et al. 2001). Enterochromaffine cells are activated by sensory neurones that monitor the pressure and chemical composition of the intestinal lumen. Reuptake of serotonin was thought to be mediated exclusively by the high affinity Na^+/serotonin cotransporter SERT. Somewhat surprisingly, the knock-out of SERT and its contribution to serotonin transport could be compensated by OCT1 and by the Na^+-dependent dopamine transporter DAT (Chen et al. 2001).

Using small specimens from human cerebral cortex we detected OCT2 protein in neurones of the human brain (Busch et al. 1998), and rOCT2 was localized by others to the apical membrane of epithelial cells of the choroid plexus (Sweet et al. 2001). In these sites, OCT2 may be involved in the uptake of choline into neurones and in choline reabsorption from the cerebrospinal fluid. Further studies of the tissue and cellular distribution of OCTs are certainly among the experimental approaches that hold the greatest promise to improve our understanding of these transporters.

Ligand specificities of OCTs

Some substrates and inhibitors have similar affinities for the different OCT subtypes in humans (Table 3), rats (Arndt et al. 2001), and mice (Kakehi et al. 2002; and H. Koepsell, unpublished data). For example, similar IC_{50} values were found for the inhibition of hOCT1, hOCT2, and hOCT3 by phenoxybenzamine, of hOCT1 and hOCT2 by procainamide, of hOCT2 and hOCT3 by metformin, and of hOCT1 and hOCT3 by β-estradiol (Table 3). Furthermore, the K_m values for MPP uptake by hOCT1 and hOCT2 are similar. In contrast, the affinity of several other substrates and inhibitors differs significantly between the OCT subtypes (Tables 3, 4). When the affinity differences between the OCT subtypes are large enough, they may be used to distinguish these subtypes experimentally. For example, an observed transport activity that might be due to either hOCT1 or hOCT2 can be assigned to one or the other subtype by exploiting the higher sensitivity of hOCT2 to a number of inhibitors, including cimetidine (ratio of respective IC_{50} values in subtypes 1 vs. 2: ~19:1), N^1-methylnicotinamide (NMN) (28:1), prostaglandin E_2 (23:1), quinine (7:1), and tetramethylammonium (70:1). The distinction between hOCT3 on the one hand and hOCT1 and/or hOCT2 on the other can be made based on the much higher sensitivity of hOCT3 to the inhibitors o-methylisoprenaline (ratio of respective IC_{50} values in subtypes 1 and 2 vs. 3: 23:1) and corticosterone (23:1).

General statements concerning the affinity of ligands to the human OCT subtypes can be made only with great caution. Thus, the "liver subtype" hOCT1 has a lower affinity for many substrates as compared to the "kidney subtype" hOCT2, including cimetidine (ratio of IC_{50} values in subtypes 1 vs. 2: 19:1), NMN (28:1), prostaglandin E_2 (23:1), quinine (7:1), tetramethylammonium (70:1), and tetrapentylammonium (5:1). Exceptions to this

Table 4 K_m and IC_{50} values that are different between the OCT subtypes from rat and mouse. The measurements were performed at physiological pH using different expression systems

Compound	rOCT1	rOCT2	rOCT3	References
	K_m or (IC_{50}) [μM]			
Rat				
Corticosterone	(151)	(4.0), (4.2)	(4.9)	Arndt et al. 2001; Wu et al. 1998b
Dopamine	19, 51	2,100, (2,300)	(384), (620)	Wu et al. 1998b; Busch et al. 1996b; Gründemann et al. 1998b
Estradiol		(85)	(1.1)	Wu et al. 1998b
Guanidine	1,660, (4,470)	172, (171)		Arndt et al. 2001
Norepinephrine		(4,400), (11,000)	(432)	Wu et al. 1998b; Gründemann et al. 1998b
o-Methylisoprenaline	(37)	(2,620)		Arndt et al. 2001
Procainamide	(20)	(445)		Arndt et al. 2001
Serotonin	38	(3,600)	(970)	Wu et al. 1998b; Busch et al. 1996b; Gründemann et al. 1998b
d-Tubericidine	23	23		Chen et al. 2002
Mouse				
	mOCT1	mOCT2	mOCT3	
Cimetidine	(0.59)	(8.0)		Kakehi et al. 2002
Procainamide	(3.9)	(312)		Kakehi et al. 2002
Quinine	(0.28)	(2.8)		Kakehi et al. 2002
o-Methylisoprenaline	(8.4)	(>100)	(1.4)	H. Koepsell et al., unpublished data

rule are β-estradiol and verapamil which have a higher affinity to hOCT1 compared to hOCT2 (1:6 and 1:70, respectively).

Table 4 shows inhibitors and substrates that may be used to distinguish OCT1, OCT2, and/or OCT3 in rat and mouse. For a given subtype of the OCT transporters, distinct species differences in affinity for substrates and inhibitors exist. For example, tetramethylammonium inhibits the MPP uptake by OCT1 with IC_{50} values of 0.9 mM in rat, of 2 mM in mouse, of 5.8 mM in rabbit, and of 12.4 mM in humans (Dresser et al. 2000). Similarly, corticosterone inhibits MPP uptake by OCT3 with IC_{50} values of 4.9 μM in rats vs. 0.12 μM in humans (Gründemann et al. 1998a; Wu et al. 1998b). Compounds that interact with two OCT subtypes may be nontransported inhibitors for one subtype and a poorly transported substrate for another. For example, quinine inhibits TEA uptake by rOCT1 and rOCT2 with IC_{50} values of 4.1 μM and 23 μM, respectively, and is transported by rOCT1 but not by rOCT2 (Arndt et al. 2001; van Montfoort et al. 2001).

Targeted disruption of OCT1 and OCT3

Knock-out mice have been generated for OCT1 and OCT3 (Jonker et al. 2001; Zwart et al. 2001). Both strains were viable and fertile, showed no obvious physiological defect and no sign for reduced embryonic viability. The effects of knocking out the rOCT1 gene on uptake, distribution, and excretion of organic cations was examined with various radioactively labelled substrates of OCT transporters. The substances were applied intravenously, and 30 min later their concentrations were determined in blood, liver, small intestine, kidney, and spleen. Four of the applied compounds (TEA, MPP, cimetidine, and choline) are common substrates of all three OCT subtypes; furthermore, TEA is also transported by OCTN1 and OCTN2. Compared to wild-type, OCT1 knock-out mice had similar tissue distributions of choline and cimetidine. In contrast, OCT1 knock-out mice showed largely reduced concentrations of TEA, MPP and *meta*-iodobenzyl-guanidine (MIBG) determined in liver, and slightly reduced concentrations in small intestine. These findings underline the importance of OCT1 for the hepatic uptake of several organic cations; furthermore, they suggest that OCT1 is involved in the small intestinal excretion of these cations. In contrast, the renal excretion of the tested cations did not depend on OCT1. More detailed experiments using TEA showed that the larger fraction of TEA was excreted by the kidney (93%) as compared to the liver (7%). Biliary excretion was reduced by 60% after disruption of OCT1. An unexpected observation was the 1.5-fold increase of urinary TEA excretion in the OCT1 knock-out mice, measured within 60 min after intravenous TEA administration. This increase could be due to a rapid upregulation of transporters that mediate renal TEA excretion such as OCT2, OCTN1, or OCTN2 (see the section entitled "Organic cation transport in liver and kidney"). As an alternative explanation, disruption of OCT1 may have destroyed the ability of the liver to act as a sink for TEA. Without that sink, higher plasma concentrations of TEA and thus enhanced renal excretion are a likely scenario. The importance of OCT1 for the uptake of organic cations into liver and small intestine was confirmed by Sugiyama and coworkers who showed that the concentration of the antidiabetic metformin 10 min after intravenous application was considerably reduced in liver and small intestine of OCT1 knock-out mice as compared to control animals; in contrast, the concentration of metformin in the kidney was not different (Wang et al. 2002).

Similar pharmacokinetic studies using radioactively labelled MPP in mice with a knock-out of OCT3 did not reveal significant changes of MPP concentration in many tissues including liver, kidney, small intestine, brain, and placenta. In contrast, the concentration of MPP was significantly reduced in the heart. These data suggest that OCT3 is not important for MPP uptake in liver, kidney and small intestine, but plays a significant role for the uptake of organic cations in heart. Interestingly, the MPP concentration in mouse embryos was significantly reduced in OCT3 knock-out mice even though the MPP concentration in the placenta was similar to controls. This may suggest that OCT3 is not necessary for the translocation of MPP and other common substrates of OCT transporters across the materno–fetal barrier of the placenta, but is essential for the uptake of organic cations in embryonic tissues where it may be expressed early during development (Slitt et al. 2002). Another explanation would be that MPP is taken up independently from OCT3 into the maternal portions of the placenta, but needs OCT3 to cross the materno–fetal barrier. Expression of OCT1 in the placenta was shown in human and rat, whereas rOCT2 could not be detected (Wessler et al. 2001a; H. Koepsell et al., unpublished data). So far, no evidence has been presented that the disruption of OCT3 leads to changes in norepinephrine concentrations in embryos.

Functional mechanism of rOCT2

In collaboration with G. Nagel (Max Planck Institute of Biophysics, Frankfurt), we carried out a series of studies focusing on the transport mechanism of rOCT2 (Arndt et al. 2001; Budiman et al. 2000; Nagel et al. 1997). The function of OCT transporters can be studied

Fig. 3a, b Choline transport by the organic cation transporter rOCT2 from rat. **a** Current-voltage relation in giant patches of *Xenopus* oocytes expressing rOCT2, inside-out configuration with symmetrical choline concentrations of 2 mM (○) or 10 mM (●); holding potential between voltage steps was 0 mV. **b** Model of a simple carrier, illustrating the individual reaction steps involved in electrogenic cation uniport (*red*) and electroneutral cation/cation exchange (*green*). T_o and T_i, transporter conformations with the cation binding site oriented outwardly and inwardly, respectively; *cat$^+$*, monovalent organic cation

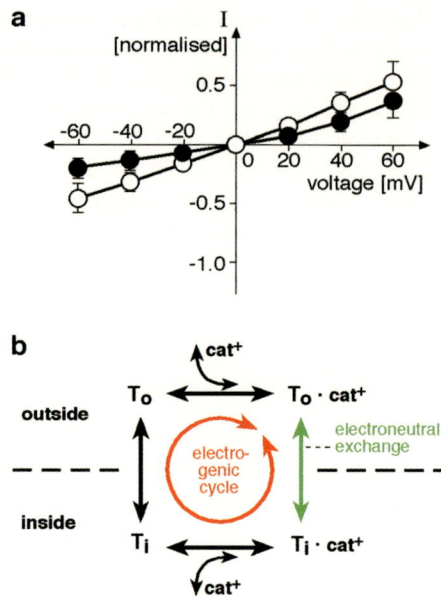

in detail by isotope uptake or voltage-clamp experiments. The electrical measurements can be performed either on intact oocytes or on giant patches excised from the oocyte plasma membrane. These methods allow short term measurements at controlled membrane potentials. Whereas in intact oocytes the intracellular concentrations of endogenous substrates and inhibitors cannot be controlled, the giant patch technique allows adjustment of the buffer composition on both membrane sides. Unfortunately, giant patch measurements are technically demanding and can be used only with highly active transporters. We selected the rat OCT2 transporter for giant patch measurements because it provided us with the highest currents as compared to other subtypes and species forms.

In the inside-out configuration, giant patches from oocytes expressing rOCT2 clamped to 0 mV yielded K_m values for the efflux of choline and TEA that were similar to the K_m values obtained at the same holding potential for the influx of these substrates in intact oocytes. This suggests symmetric translocation of small cations by rOCT2. When the membrane potential was set to values different from 0 mV, the effects on the K_m values of choline and TEA were different for the influx as compared to the efflux. This suggests that the binding or debinding of cations to or from rOCT2 occurs within the electrical field of the plasma membrane. Because rOCT2 operates in a rather symmetric way and mediates the translocation of many organic cations, we tried to determine whether rOCT2 conforms to the simple alternating access model according to which the translocation of substrate is stoichiometrically linked to conformational changes of the transporter, or whether it may also function as a channel. Figure 3a shows recordings from giant patches with symmetrical choline concentrations and at various membrane potentials. The experiments demonstrate that both inward and outward currents are voltage dependent in a way that is rather symmetric around the reversal potential.

Interestingly, considerably higher currents were observed with 2 mM choline as compared to 10 mM choline on both sides of the membrane. Such a result would be expected for the simple alternating access model: in addition to electrogenic transport as the result of the sequence of steps (substrate binding to the outside → translocation of the bound substrate to the intracellular side → debinding of substrate → reorientation of the empty substrate binding site towards the extracellular side), this model also predicts electroneutral cation exchange at high substrate concentrations (Fig. 3b). Alternatively, a channel with an inhibitory, low-affinity cation binding site may show a similar decrease of current at high substrate concentrations. However, we could rule out a channel-like transport mode by demonstrating *trans*-stimulation of dopamine efflux by choline under voltage-clamp conditions (G. Nagel, H. Koepsell et al., unpublished data). *Trans*-stimulation of a transporter is expected if the in-to-out reorientation of the substrate binding site of the loaded carrier occurs more quickly than reorientation of the empty one. A channel may show *trans*-stimulation only if it contains a *trans*-stimulating cation binding site. Thus, the notion that rOCT2 is a channel can only be maintained together with the additional assumptions of both a stimulatory extracellular cation binding site (to explain *trans*-stimulation by bath choline) and a second, inhibitory cation binding site that is somehow able to override the effect of the stimulatory site (to explain the reduction of currents at high substrate concentrations under symmetric conditions). Because these assumptions are not supported by any independent evidence and lead to rather complicated models, whereas the same observations can be explained easily with the simple carrier, these experiments strongly support the idea that rOCT2 is a transporter rather than a channel.

The substrate binding sites of OCT transporters

We recently characterized the outwardly and inwardly directed conformations of the substrate binding sites of rOCT2 using the nontransported inhibitors tetrabutylammonium and corticosterone (C. Volk and H. Koepsell, unpublished data). These inhibitors were used to measure the short-term inhibition of inward currents in intact oocytes induced by application of choline from the outside, and outward currents in giant patches induced by choline application from the inside. The experiments suggest that either inhibitor interacts both with the inwardly directed and outwardly directed conformations of the substrate binding sites. The inhibition of choline transport by tetrabutylammonium and corticosterone could be completely or partially overcome by high choline concentrations. Interestingly, tetrabutylammonium had a significantly higher affinity from the outside compared to the inside, whereas corticosterone had a higher affinity from the inside compared to the outside. The data suggest that the polyspecific OCT transporters contain binding pockets with partially overlapping interaction domains for different substrates and inhibitors. Reorientation of the substrate binding pockets may be accompanied by structural changes of the binding pocket that alter the affinity for substrates or inhibitors. As a consequence, some inhibitors or substrates may have different affinities on the outside vs. the inside. Because the different ligands have different interaction sites, those affinity changes may go in different directions for different ligands. It is thus easy to imagine that competition between inhibitors and substrates within the binding pocket may vary considerably between individual inhibitors or substrates, and may be different from outside than from inside.

Regulation of OCTs

The cloning of the polyspecific organic cation transporters made it possible to address the regulation on the level of individual OCT subtypes. Knowledge about the specific regulatory pathways may help to devise pharmacotherapeutic strategies. OCTs appear to be regulated both on the mRNA and protein level. In addition, regulatory processes may alter the functional properties of OCT transporters within the plasma membrane. Furthermore, the available data suggest that the regulation of OCTs is subtype-, tissue-, and species-dependent.

Long-term regulation is evidenced by expression of cation transport in rat kidney that is gender dependent and changed during diabetes (Bowman and Hook 1972; Grover et al. 2002). In male rats, basolateral uptake of TEA into renal proximal tubule cells was twice as high as in female rats and correlated with increased levels of rOCT2 mRNA and protein (Urakami et al. 1999). Application of testosterone to female rats resulted in increased expression of OCT2, whereas application of estradiol to male rats decreased the expression of OCT2 (Urakami et al. 2000). Levels of OCT2 mRNA in Mardin Darby canine kidney (MDCK) cells increased about twofold following exposure to dexamethasone, hydrocortisone or testosterone (Shu et al. 2001). In rats with streptozotocin-induced diabetes, TEA transport into sections of rat kidney cortex was diminished, and this effect could be reversed by administration of insulin (Grover et al. 2002). Sequencing of the promoter regions of the human OCT2 and OCT3 genes revealed consensus sequences for the binding of several ubiquitous and specific transcription factors (Gründemann and Schömig 2000), but the binding and/or functional activity of the respective transcription factors was not studied.

Short-term regulation of rat and human OCT1, OCT2 and/or OCT3 has been investigated in HEK 293 cells and in *Xenopus* oocytes (Cetinkaya et al. 2003; Martel et al. 2001; Mehrens et al. 2000; Schlatter et al. 2002). Some additional experiments were performed in isolated proximal tubules from rabbit and human kidney (Hohage et al. 1994; Pietig et al. 2001). In HEK 293 cells, rOCT1-mediated transport of 4-(4-(dimethylamino)styryl)-*N*-methylpyridinium iodide (ASP) was stimulated by activators of protein kinase C (PKC), of protein kinase A (PKA), and of tyrosine kinase (Mehrens et al. 2000), but inhibited by cGMP (Schlatter et al. 2002). In HEK 293 cells, PKC-dependent stimulation of rOCT1 was accompanied by phosphorylation of rOCT1, whereas inhibition of rOCT1 by cGMP appeared to be independent from phosphorylation (Mehrens et al. 2000; Schlatter et al. 2002). Since the inhibition of rOCT1 by cGMP was only observed in HEK 293 cells but not in *Xenopus* oocytes, the effect of cGMP might require a regulatory protein that is not expressed in the oocytes. Interestingly, substrate selectivity of rOCT1 was changed after stimulation of PKC by sn-1,2-dioctanoyl glycerol (DOG; Mehrens et al. 2000); the affinities of TEA, tetrapentylammonium, and quinine for rOCT1 were increased by factors of 58, 15, and 2, respectively. At variance to rOCT1, uptake of ASP into HEK-293 cells stably expressing hOCT2 was not affected after stimulation of PKC by DOG, and was slightly inhibited after stimulation of PKA by forskolin (Cetinkaya et al. 2003). Furthermore, hOCT2 was inhibited by the muscarinic receptor agonist carbachol that is known to increase intracellular Ca^{2+} concentration via an activation of the phospholipase C (PLC) pathway. ASP uptake by hOCT2 was largely reduced by calmidazolium which inhibits the Ca^{2+}/calmodulin complex (Ca^{2+}/CaM), by inhibitors of the Ca^{2+}/CaM dependent protein kinase II, or by inhibitors of Ca^{2+}/CaM dependent myosin light-chain kinase. This indicates that OCT2 is constitutively activated by the Ca^{2+}/calmodulin complex. The effect of carbachol, however, appears to be independent from Ca^{2+}/CaM because it was not sensitive to calmidazolium. On the other hand, inhibition of hOCT2 by carbachol was abolished after blocking of phosphatidyl 3-kinase, another pathway downstream to PLC. Similarly to rOCT1, regulation might affect affinity of hOCT2: inhibition of the $Ca^{2++}/calmodulin$ complex by calmidazolium decreased the affinity of hOCT2 for TEA tenfold (Cetinkaya et al. 2003).

To compare the regulation of hOCT2 in heterologous expression systems with that prevailing in native tissues, uptake of ASP was measured across the basolateral membrane of isolated human proximal tubules (Pietig et al. 2001) which is likely to reflect hOCT2 activity based on the known localization of hOCT2 in the basolateral membrane of renal proximal tubules (Motohashi et al. 2002) and the typical inhibition characteristics (Pietig et al. 2001). Given the low levels of OCT1 in human kidney (Gorboulev et al. 1997) and the highly reduced activity of the hOCT2 splice variant described above (Urakami et al. 2002), ASP uptake via transporters other than hOCT2 is expected to be less than 10% of hOCT2-mediated uptake.

Basolateral uptake of ASP in human proximal tubules was inhibited upon activation of PKA by forskolin, in keeping with the findings in HEK 293 cells expressing hOCT2. Stimulation of PKC, however, inhibited ASP uptake in the proximal tubules, at variance to hOCT2-expressed uptake in HEK 293 cells, suggesting that the regulation of hOCT2 is strongly modulated by the differential expression of signaling pathways or cofactors in different cell types.

The regulation of human hOCT3 was examined after stable expression in HEK 293 cells (Martel et al. 2001). Similar to hOCT2, cation uptake mediated by hOCT3 was reduced by inhibition of the $Ca^{2+}/calmodulin$ complex. Unlike hOCT2, which was inactivated by inhibition of Ca^{2+}/CaM-dependent kinases, hOCT3 was inactivated by inhibition of a Ca^{2+}/CaM-dependent phosphodiesterase, PDE1. Furthermore, hOCT3 was not significantly affected by PKA.

Taken together, the data indicate complex subtype specific short-term regulations of the organic cation transporters. The regulations may involve transporter phosphorylation and regulatory proteins; both may alter the functional properties of the transporter as well as transporter concentrations within the plasma membrane. Apart from phosphorylation and regulatory proteins, the observed effects might be explained by indirect effects, for example by changes affecting membrane potential or the concentration of endogenous organic cations or anions (Goralski et al. 2002).

Structure–function relations in OCTs: insights from mutagenesis studies

Amino acids that are common to the OCTs but different in OAT and/or OCTN transporters provide a reasonable point of departure in attempts to identify individual amino acids of rOCT1 that are essential for organic cation transport. Of these, a change of Asp145 in the large extracellular loop (Fig. 4) to histidine had no effect on cation transport (Chen et al. 2002). In contrast, mutations of Asp475 in the middle of the predicted 11th transmembrane domain (see Figs. 1, 4) changed transport selectivity (Gorboulev et al. 1999); OATs 1–5, URAT1, OCTN1, and OCTN2 all carry an arginine rather than aspartate in this position. When we replaced Asp475 of rOCT1 by arginine, asparagine, or glutamate, we observed specific transport rates in HEK-293 cells that were less than 10% as compared to wild type. In *Xenopus* oocytes, however, only the Asp475Glu mutant yielded detectable

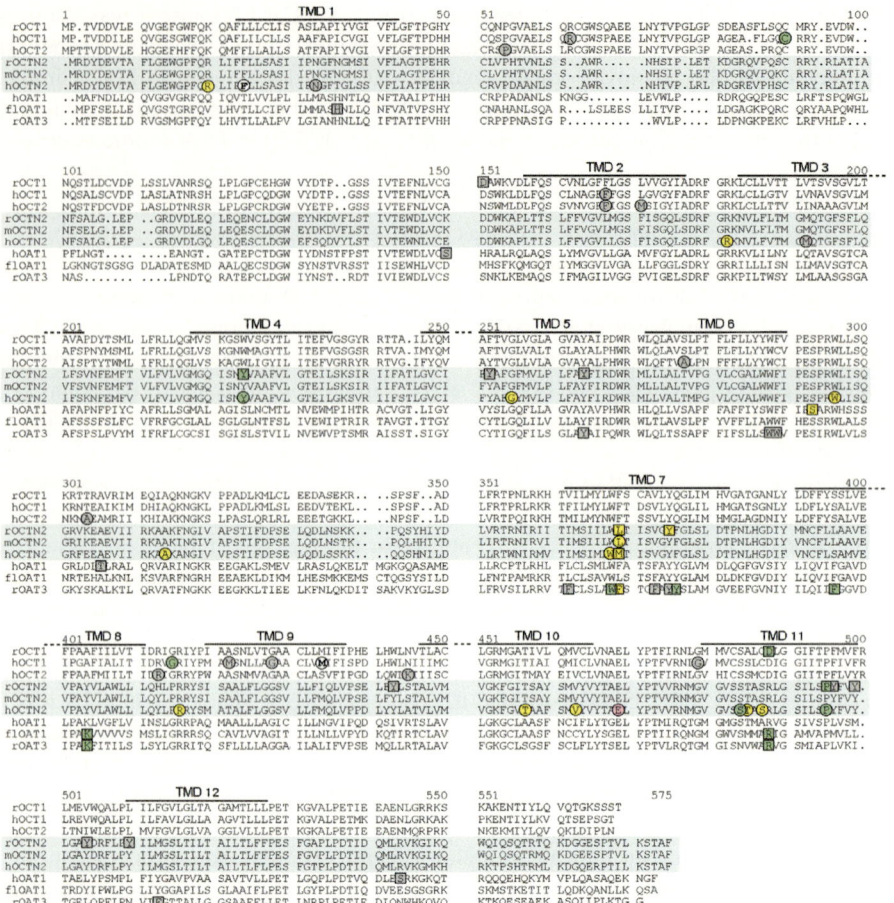

Fig. 4 Amino acid sequence alignment of selected SLC22 family members, indicating the position of spontaneous or genetically engineered mutations that elicit specific changes of transport function. The alignment was performed with the GCG programs PileUp and LineUp (Wisconsin Sequence Analysis Package, version 8.0. Genetic Computer Group. University Research Park, Madison, Wisconsin, 2000). Abbreviated gene names as in Table 2 and in the text. References are given in the text. *Boxed* amino acids, sites of point mutations engineered in vitro; *circled* a.a., sites of spontaneous point mutations identified in the respective species; *gray,* a.a. whose mutation did not significantly affect transport; *white,* amino acids whose deletion did not significantly affect transport; *yellow,* sites of missense mutations that reduced transport activity to <10% of wild type when assayed in fibrocytes of homozygous patients or in heterologous expression systems; *green,* sites of missense mutations that caused specific changes of substrate affinity and/or substrate selectivity; *red,* site of missense mutation in hOCTN2 that led to a reduced Na$^+$-affinity of Na$^+$/L-carnitine cotransport

transport, which was reduced to 2.3% for TEA, 3.2% for NMN, 3.5% for choline, and 11.4% for MPP.

Interestingly, the Asp475Glu mutant exhibited considerably higher apparent affinities for TEA, NMN, and choline than wild-type OCT1 (respective K_m values reduced by factors of 8, 3.5, and 15), whereas affinity for MPP was unchanged. Similarly, inhibition of TEA uptake by the Asp475Glu mutant occurred with affinities that were higher for some inhibitors, but unchanged for others. Two tentative conclusions were drawn from these observations: firstly, Asp475 is probably close to the substrate binding site of rOCT1; secondly, the cation binding site of rOCT1 behaves as a pocket that offers several, only partially overlapping interaction domains for different substrates. Such a crucial role of Asp475 in the 11th potential transmembrane domain rOCT1 seems to be at odds with the reported transport activity of a hOCT2 splice variant that lacked the entire C-terminus including the tenth, 11th and 12th transmembrane domain (Urakami et al. 2002). Because the essential role of the 11th TMD is well established for further members of the SLC22 family, we hypothesize that the residual activity observed with this splice variant was either due to an activation of an endogenous transport activity, or to a second, alternate transport path within the same transporter molecule. Thus, point mutations of the arginine residues in the organic anion transporters OAT1 from flounder (fOAT1) and OAT3 from rat (rOAT3) that correspond to Asp475 in rOCT1 (see Fig. 4) produced similar specific functional changes (Feng et al. 2001; Wolff et al. 2001). In fOAT1, the mutation Arg478Asp decreased affinity and maximal transport rate for para-amminohippurate (PAH; Wolff et al. 2001), and abolished interaction with glutarate. In rOAT3, the mutations Arg454Asp and Arg454Asn (see Fig. 4) both changed substrate selectivity with respect to the anions PAH and ochratoxin A, the weak base cimetidine and the permanently charged cation MPP (Feng et al. 2001). The pivotal functional importance of the predicted 11th TMD was further confirmed for OCTN2 by point mutations in other positions of the 11th TMDs (see the section entitled "The carnitine and cation transporters of the SLC22 family" and Fig. 4).

Other data suggest that the eighth transmembrane domain may also be part of the substrate binding pocket in transporters of the SLC22 family. The mutation Lys370 in the eighth transmembrane domain of rOAT3 to alanine changed substrate selectivity; for example, uptake of PAH was reduced to a considerably higher extent than the uptake of cimetidine (Feng et al. 2001). Moreover, substrate selectivity of fOAT1 was changed when Lys394 (corresponding to Lys370 of rOAT3) was mutated to alanine (Fig. 4): *trans*-stimulation of PAH efflux by extracellular glutarate was abolished, whereas *trans*-stimulation of PAH efflux by extracellular PAH remained functional (Wolff et al. 2001). In rOAT3, an additive effect on substrate recognition was observed in a comparison of MPP uptake in the double mutant K370A/R454D vs. the R454D mutant.

Finally, recent data from our laboratory show an involvement of the fourth transmembrane domain of rOCT1 in substrate recognition. Mutations of two amino acids in the fourth transmembrane domain selectively and significantly increased substrate affinities for certain substrates. Specific effects on transport were also observed after mutation of other amino acids in the presumed fourth TMD of human OCTN2 (see the section entitled "The carnitine and cation transporters of the SLC22 family" and Fig. 4).

The functional role of the large extracellular loop connecting TMD-1 and 2 (ECL-1,2) of the SLC22 transporters is not well understood. In rOCT1, ECL-1,2 is not required for transport, but can affect it nonetheless. Thus, on the one hand, organic cation transport was preserved in a splice variant of rOCT1 lacking the N-terminus including TMD-1, TMD-2, and ECL-1,2 (Zhang et al. 1997b). Replacing ECL-1,2 of rOCT1 by ECL-1,2 of rOCT2 did not affect the IC_{50} of rOCT1 for several cations that have largely different affinities in rOCT2 (H. Koepsell et al., unpublished data). On the other hand, several point mutations within the large ECL-1,2 inactivate rOCT1. Furthermore, mutation of Cys88 in the ECL-1,2 of hOCT1 drastically decreased transport rates and altered the substrate selectivity (Kerb et al. 2002).

In conclusion, in-vitro mutagenesis studies in organic anion and cation transporters of the SCL22 family indicate that transmembrane domains 4, 8, and 10, possibly together with additional domains, determine the structure of the substrate binding pocket, either by providing sites of direct interaction with substrates, or by indirectly stabilizing the confor-mation of those binding sites. The large extracellular loop may indirectly contribute to the formation or stabilization of the substrate binding pocket. No data are available that allow any insight into the translocation mechanism. The interpretation of the mutations is limited by the fact that the membrane topology of these transporters has not been determined bio-chemically and that it is not known whether these transporters operate as monomers, di-mers, or oligomers.

Polymorphisms in human OCT1 and OCT2

A number of mutations and polymorphisms have been identified recently in the human OCT1 and OCT2 genes (Kerb et al. 2002; Leabman et al. 2002). Since OCT1 and OCT2 are critically involved in the hepatic and renal excretion of cationic drugs, functionally rel-evant mutations in these transporters can alter the handling of certain drugs in the body and thus enhance (or decrease, depending on the particular situation) toxicity and thera-peutic efficacy. In a population of 57 Caucasians, numerous single nucleotide polymor-phisms (SNPs) within the hOCT1 gene were detected and further analyzed. Eight SNPs resulted in single amino acid substitutions (Fig. 4). Out of these, three SNPs affected or-ganic cation transport in vitro. Two SNPs are located in the large extracellular loop (Arg61Cys and Cys88Arg), and one in the short intracellular loop between the TMD-8 and TMD-9 (Gly401Ser). The frequencies of genotypes heterozygous for these mutants were 16% (Arg61Cys), 1.2% (Cys88Arg), and 6.5% (Gly401Ser). Uptake of 0.1 µM MPP uptake by Arg61Cys, the most frequent mutant, was about 50% compared to wild-type hOCT1, with no differences in affinity. The less frequent mutants Cys88Arg and Gly401-Ser exhibited MPP uptake of less than 2% of wild-type hOCT1. Interestingly, the substrate selectivity of these mutants was changed. Since hOCT1 is required for hepatic uptake of cationic drugs and xenobiotics, the mutations with reduced transport activity and/or chan-ged substrate selectivity are supposed to affect the hepatic excretion and toxic disposition of certain drugs and xenobiotics, for better or worse.

The occurrence of SNPs in the gene coding for hOCT2 was investigated in 247 individ-uals of various ethnicity (Leabman et al. 2002). This study identified 28 variable sites, 16 of which were localized in coding regions. Eight of these caused single amino acid substi-tutions at seven positions (Fig. 4), and one caused a premature termination of the protein. Met165Ile and Arg400Cys were only observed in African-Americans, and Lys432Gln only in African-Americans and Mexican-Americans. Furthermore, the ratio of synony-mous over nonsynonymous nucleotide changes was significantly higher than the reported genetic variations in a population of more than 75 other genes (Cargill et al. 1999; Halush-ka et al. 1999). The higher degree of amino acid conservation in hOCT2 implicates a high-er selective pressure, underlining the biological importance of hOCT2, which is probably related to the elimination and detoxification of organic cations.

This view is further supported by the observation that a gain of function was found in three out of four tested frequent, nonsynonymous SNPs (Ala270Ser, Met165Ile, Arg400Cys, and Lys432Gln). Mutants Met165Ile and Arg400Cys took up MPP at saturat-ing concentrations at a rate 2–3 times higher than wild-type hOCT2, and Lys432Gln showed apparent K_m values for MPP and tetrabutylammonium that were reduced by 44% and increased by 48%, respectively. Future studies may show whether the rare mutations are associated with altered function. Studies in individuals carrying the identified gain-of-function mutations as well as carriers of other, hitherto unknown mutations of the hOCT2 gene may help to elucidate the impact of hOCTs to drug distribution and action, and might provide a rationale for improving drug efficacy or reducing drug toxicity.

Clinical implications of OCTs

OCTs are involved in a variety of physiological and pathophysiological processes. OCT1 and OCT2 mediate the first step in hepatic or renal excretion of many cationic and amphiphilic drugs. Moreover, OCT1 appears to be involved in the absorption of drugs across the epithelium of the small intestine. All OCTs affect the interstitial concentrations of endogenous compounds (e.g., choline and monoamine neurotransmitters), drugs, and xenobiotics in a variety of tissues including brain and heart. OCT1 and OCT3 may mediate the cellular release of acetylcholine from the placenta during nonneuronal cholinergic regulation (Wessler et al. 1999, 2001b). So far, no specific disease or adverse drug reaction was attributed to any of the OCTs. This is not astonishing because the OCT subtypes have overlapping expression patterns and substrate specificities, and polyspecific transporters from other families share OCT substrates (see the section entitled "Overlap between substrate specificities of OCTs and/or OCTNs with other polyspecific transporters") so that defects in function or expression of OCTs may manifest only upon exposure to certain drugs or xenobiotics. For instance, some drugs might be less promiscuous than others with respect to transporters, and comedication may inhibit alternative transport pathways. The identification of functionally relevant mutations in OCT1 is the first step to investigate this issue.

Mutations in OCT1 or proteins that are involved in its targeting or membrane turnover may result in reduced hepatic excretion of OCT1 substrates (including drugs) that are eliminated mainly via biliary excretion. The resulting higher-than-normal plasma levels may make similar therapeutic effects possible with lower dosage, or may cause untoward side effects. Low expression of hOCT2 in kidney, or defect mutations in hOCT2 potentially related to one of four uncharacterized mutations (Leabman et al. 2002) may reduce renal excretion of more hydrophilic cationic drugs. For drugs transported by both OCT1 and OCT2, reduced function of OCT2 may lead to increased hepatic elimination or toxicity.

Comedication with drugs that are substrates or inhibitors of OCTs may have severe consequences. For example, the renal and/or hepatic excretion of a drug that is translocated by OCT1 and/or OCT2 will be impeded by comedication that blocks both OCT1 and OCT2. At variance, hepatic excretion will be increased by comedication with blockers specific for OCT2, but not OCT1. In this context it should be noted that weak bases of hydrophobic compounds may inhibit OCT transporters from the cytoplasmic side and may not be removed easily (Arndt et al. 2001; H. Koepsell et al., unpublished data) and that the interaction of two specific drugs at the outwardly or inwardly directed binding pocket of OCT1 or OCT2 cannot be predicted. Substrates and inhibitors may compete by different degrees and their interaction from intracellular or extracellular may be different.

In rat, steroid hormones influence the expression of OCT2 (see above). Inui and coworkers showed that the reduced excretion of the organic cation cimetidine in 5/6 nephrectomized rats was associated with a 50% decrease of plasma testosterone and downregulation of OCT2 (Ji et al. 2002); the expression of OCT1 and OAT3 was not changed.

Downregulation of OCT2 after nephrectomy was reverted upon intravenous infusion of testosterone. These findings raise the attractive and testable hypothesis that impaired renal excretion of cationic drugs in patients suffering from chronic renal failure may as well be improved by testosterone. Conversely, orchidectomia (e.g., in patients suffering from prostate cancer) may be associated with an increased risk for accumulation of cationic drugs.

Two examples of how substrates of OCT transporters may produce adverse drug effects if their transport by OCTs is impaired or increased are worth noting. It has been reported that the treatment of peptic ulcer with the H_2 histamine receptor blocker cimetidine leads to mental confusion in some patients that is reversible after withdrawal of cimetidine (Kimelblatt et al. 1980; Schentag et al. 1979). Since this drug effect was correlated with increased levels of cimetidine in plasma and cerebrospinal fluid, and cimetidine is transported by OCT1 and OCT2, decreased expression or malfunction of these transporters could be the reason for increased cimetidine concentrations and the associated neurological symptoms. Another example is lactic acidosis, which may occur as a rare but life-threatening side effect in type-II diabetics who are treated with biguanides such as phenformin and metformin (Davidson and Peters 1997). Because of that risk, phenformin was withdrawn from the market in the 1970s (Kwong and Brubacher 1998); metformin, however, is still used. In recent years, metformin has also been introduced for treatment of polycystic ovary syndrome (Nestler 2001; Velazquez et al. 1994). Metformin is mainly eliminated by glomerular filtration and tubular secretion, but part of it may be also excreted into the bile. Its antidiabetic effect is due to improved peripheral insulin sensitivity, reduced glucose absorption in small intestine, and reduced glucose generation by the liver (Borst and Snellen 2001; Caspary and Creutzfeldt 1971; Hundal et al. 2000). The effect of metformin on glucose generation in hepatocytes was explained by inhibition of mitochondrial complex I (El Mir et al. 2000; Owen et al. 2000). Extensive inhibition of mitochondrial respiration by biguanides may cause lactic acidosis. Renal failure or impaired renal excretion of metformin may entail increased metformin plasma levels and thus cause lactic acidosis. Metformin is transported by OCT1 in rat and humans, and by OCT2 in humans (Dresser et al. 2002; Wang et al. 2002). In humans, lower renal expression of OCT2, mutations reducing the activity of hOCT2, or administration of metformin simultaneously with a drug that inhibits OCT2 in kidney but not hOCT1 in the liver, may cause lactic acidosis.

OCT transporters could play a role in carcinogenesis or may be useful for targeting of anticancer drugs. For example, high expression or activity of OCT1 may increase the concentration of carcinogenic xenobiotics in hepatocytes. On the other hand, high expression of OCT1 was found in hepatocarcinomas induced by diethylnitrosamine (Lecureur et al. 1998), and thus OCT1 could help target anticancer drugs into the carcinoma cells. To prevent renal and hepatic toxicity, the drug should not interact with OCT2 and OCT3, and OCT1 expressed in nontransformed hepatocytes should be downregulated. This could be achieved by employing differences in the regulation OCT1 in normal compared to transformed liver cells. Bile acid-conjugates of the anticancer drug cisplatin were recently shown to be transported substrates of human OCT1 and OCT2, in contrast to free cisplatin (Briz et al. 2002). These drugs may not be suitable for targeting because they are also transported by at least three other transporters expressed in liver (see the section entitled "Organic cation transport in liver and kidney").

The carnitine and cation transporters of the SLC22 family

OCTN1

Homology screening led to the identification of four members of the SLC22 family that translocate the zwitterion carnitine together with Na^+ and/or organic cations (hOCTN1 und hOCTN2 in Fig. 2, Tables 2, 3). OCTN1 was cloned from human, rat, and mouse (Tamai et al. 1997, 2000; Wu et al. 2000a; Yabuuchi et al. 1999). In human, OCTN1 is expressed in kidney, skeletal muscle, placenta, prostate, heart, and lung (Tamai et al. 1997).

Human OCTN1 (hOCTN1) transports the cations TEA, quinidine, pyrilamine, and verapamil and the zwitterion carnitine (Yabuuchi et al. 1999). Many other cations interact with hOCTN1, including choline, cimetidine, clonidine, nicotine, procainamide, quinine, the zwitterion cephaloridine, and the anions ofloxacin and levofloxacin. The data show that hOCTN1 is a polyspecific transporter that may have a preference for organic cations. hOCTN1 mediates TEA transport in either direction with an apparent K_m of 0.2–0.4 mM in the influx mode (Tamai et al. 1997; Yabuuchi et al. 1999). Since TEA uptake by hOCTN1 is independent from an inwardly directed Na^+-gradient and from the membrane potential whereas an inwardly directed H^+-gradient stimulates TEA efflux (Yabuuchi et al. 1999), hOCTN1 may work as an electroneutral H^+/organic cation antiporter that mediates cellular efflux of organic cations.

Such electroneutral H^+/organic cation exchange activity is present in brush-border membranes of renal proximal tubules from different species (Inui et al. 1985; Jung et al. 1989; Kinsella et al. 1979; McKinney and Kunnemann 1985; Ott et al. 1991; Sokol et al. 1985; Takano et al. 1984; Wright 1985). We speculate that OCTN1 has a luminal location and is engaged in cation secretion in the proximal tubule. However, note that expression of hOCTN1 in renal brush border membranes does not fully explain uptake experiments on human renal brush border membrane vesicles because these showed electroneutral H^+/organic cation exchange not only for TEA, but also for NMN (Ott et al. 1991) whereas hOCTN1 has a very low affinity for NMN. Large species differences exist with regard to localization and function of OCTN1. For example, OCTN1 could not be detected in liver from adult humans, whereas in rats, the liver is the site where hOCTN1 is expressed most strongly (Wu et al. 2000a). Also, carnitine uptake by OCTN1 was Na^+-dependent for mouse OCTN1, but not for rat OCTN1 (Tamai et al. 2000; Yabuuchi et al. 1999).

OCTN2, OCTN3 and CT2

OCTN2 has been cloned from human, mouse, and rat (Brooks and Krähenbühl 2001; Nezu et al. 1999; Schömig et al. 1998; Sekine et al. 1998; Tamai et al. 1998, 2000; Wu et al. 1998a). In humans, this transporter is widely expressed in tissues such as skeletal muscle, kidney, prostate, lung, pancreas, heart, small intestine, adrenal gland, thyroid, and some regions of the brain (Durán et al. 2002; Schömig et al. 1998; Sekine et al. 1998; Tamai et al. 1998; Wu et al. 1998a). In mouse and rat, OCTN2 could be localized to the apical membrane of renal tubular epithelial cells (Tamai et al. 2001). OCTN2 is a Na^+/carnitine cotransporter with a high affinity for carnitine, but can function alternatively as a polyspecific and Na^+-independent cation uniporter (Tamai et al. 1998, 2001; Wagner et al. 2000;

Wu et al. 1999). Na^+-dependent transport of L-carnitine by OCTN2 is electrogenic and to some degree stereospecific (Ohashi et al. 1999; Wagner et al. 2000).

Uptake measurements with plasma membrane vesicles isolated from HEK-293 cells stably expressing human OCTN2 showed that an inwardly directed Na^+ gradient could bring about an "overshoot," i.e., the transient increase of intravesicular L-carnitine concentration beyond the final equilibrium value (Tamai et al. 2001). This phenomenon is due to the gradual dissipation of the Na^+ gradient and provides good evidence, together with the demonstration of electrogenicity, that hOCTN2 is a Na^+-L-carnitine cotransporter.

In uptake measurements, human OCTN2 exhibited an apparent K_m for L-carnitine of 4–5 µM (Tamai et al. 1998; Wagner et al. 2000), and half-maximal concentrations for Na^+-activation between 2 and 19 mM (Ohashi et al. 1999; Seth et al. 1999; Wagner et al. 2000). Thus, OCTN2 is likely to represent the molecule responsible for the high-affinity Na^+/L-carnitine cotransport activity observed in plasma membrane vesicles from skeletal muscle and in brush-border membrane vesicles from kidney and intestine (Berardi et al. 2000; Prasad et al. 1996; Rebouche and Mack 1984; Roque et al. 1996; Stieger et al. 1995).

In the presence of Na^+, hOCTN2 also transports short-chain acyl esters of carnitine with high affinity. For example, acetyl-L-carnitine is translocated by hOCTN2 with a K_m of ~8.5 µM (Ohashi et al. 1999). Acetyl-L-carnitine is of biomedical interest since it protects from neurotoxic effects of ammonium (Matsuoka and Igisu 1993), MPP, or 1-methyl-4-phenyl-1,2,3,6-tetrahydropyridine (MPTP; Harik and Hritz 1993; Steffen et al. 1995). Furthermore, acetyl-L-carnitine was tested as a treatment for Alzheimer's disease (Pettegrew et al. 1995). In the presence of Na^+, hOCTN2 can also transport cephaloridine, a zwitterionic β-lactam antibiotic (Ganapathy et al. 2000), as well as probably L-lysine and L-methionine, as suggested by electrophysiological assays (Wagner et al. 2000).

Independently from Na^+, OCTN2 from human, rat, and mouse are capable of translocating or interacting with various organic cations (Ohashi et al. 2001; Wu et al. 1999). Thus, hOCTN2 translocates the cations TEA, pyrilamine, quinidine, verapamil, and choline (Ohashi et al. 1999; Wagner et al. 2000). For TEA uptake by hOCTN2, a K_m of ~0.3 mM was determined (Ohashi et al. 2002). Carnitine uptake by hOCTN2 was competitively inhibited by TEA, and TEA uptake by hOCTN2 was competitively inhibited by carnitine (Ohashi et al. 2002; Seth et al. 1999). Inhibition of L-carnitine uptake via OCTN2 in the presence of Na^+ was found for other organic cations or weak bases (e.g., nicotine, MPP, MPTP, procainamide, quinine, metamphetamine, emetine, clonidine, and cimetidine), zwitterions (cephaloridine, cefepime, and cefoselis) and noncharged compounds (e.g., corticosterone and aldosterone; Ganapathy et al. 2000; Ohashi et al. 1999; Wagner et al. 2000; Wu et al. 1998a). The IC_{50} values for cephaloridine, cefepime, and cefoselis were 0.23, 1.7, and 6.4 mM, respectively. Interestingly, Na^+ removal decreased the affinity for cephaloridine (~tenfold increase of K_m), but not for cefepime and cefoselis (Ganapathy et al. 2000). These data suggest that binding of carnitine and cations occurs within one binding pocket.

The dual function of OCTN2 as a Na^+-driven cotransporter for certain zwitterions on the one hand, and as a Na^+-independent transporter of organic cations on the other is of great physiologic and theoretical interest. It thus appears possible that the electrogenic OCTs can similarly function as Na^+-driven cotransporters for hitherto unknown solutes. Moreover, the question arises whether Na^+ binds within the same binding pocket as the organic cations, and whether translocation of Na^+ occurs along the same transport path as cations and/or L-carnitine; in the Na^+-D-glucose cotransporter SGLT1, separate transport pathways in different parts of the molecule have been proposed for Na^+ and glucose (Wright et al. 1998).

Because OCTN2 is present in the brush border membrane of renal proximal tubules in rat and mouse (Tamai et al. 2001), and mouse OCTN2 can translocate organic cations in either directions, and there is *trans*-stimulation of carnitine uptake by TEA as well as of TEA efflux by carnitine (Ohashi et al. 2001), it is likely that OCTN2 participates in absorption and secretion of organic cations in small intestine and kidney (Fig. 6). Tsuji and coworkers demonstrated significant contribution of OCTN2 to the renal excretion of TEA by measuring the excretion and tissue distribution of intravenously injected ^{14}C TEA in control mice and juvenile visceral steatosis (*jvs*)-mice that suffer from a homozygous loss-of-function mutation of OCTN2 (Hashimoto et al. 1998; Lu et al. 1998; Nezu et al. 1999; Ohashi et al. 2001; Yokogawa et al. 1999). Secretion of TEA was reduced by 50% in *jvs* mice as compared to controls, and accumulation of TEA in the kidney was increased by 150% (Ohashi et al. 2001). Similar studies in healthy and affected humans might help to elucidate the role of OCTN2 in the human physiology of organic cation handling.

The transporters OCTN3 and CT2 belong to the same subgroup of the SLC22 family as OCTN1 and OCTN2 (Fig. 2). OCTN3 was found only in mice (Tamai et al. 2000) where it is expressed in testis and kidney. Mouse OCTN3 translocates L-carnitine with an apparent K_m of 3 µM, i.e., with an affinity seven times higher than mouse OCTN2. In contrast to OCTN2 from different species including mouse, OCTN3 could transport carnitine independently from Na^+. Given that carnitine uptake by OCTN3 was not inhibited by 0.5 mM choline, and inhibited only by 54% by 0.5 mM TEA, OCTN3 appears less relevant quantitatively for organic cation transport than OCTN1 and OCTN2.

CT2 is another high-affinity carnitine transporter that has been identified in human (Enomoto et al. 2002b). Apart from several amino acids in the N-terminal part of the protein, CT2 is identical to a gene product named OCT6 found in hematopoietic cells (Gong et al. 2002). CT2 is expressed in the testis by Sertoli cells, and by epithelial cells of the epididymal ducts (Enomoto et al. 2002b). CT2 mRNA is further expressed in fetal liver, bone marrow, leukocytes, and some leukemic cell lines (Gong et al. 2002). CT2 has a higher substrate selectivity than OCT1–3 and OCTN1 and -2 insofar as it interacts with carnitine and betaine, but not with TEA and several organic anions (Enomoto et al. 2002b). CT2 can translocate carnitine in either direction across the plasma membrane and may thus contribute to the transepithelial transport of carnitine in the epididymis.

Systemic carnitine deficiency and mutations in OCTN2

"Primary systemic carnitine deficiency" (SCD) is a recessively inherited disorder of mitochondrial fatty acid oxidation that is caused by a defect of cellular carnitine uptake (Karpati et al. 1975; Kerner and Hoppel 1998; Tein et al. 1990; Treem et al. 1988). Because intracellular carnitine is required for transferring long-chain fatty acids from the cytosol into the micochondria, the lack of carnitine interferes with the ability to fuel metabolism via the oxidation of fatty acids; this ability is of particular importance during fasting and stress. Early in life, SCD may cause acute hypoketotic hypoglycemia, Reye syndrome (encephalopathy with hyperammonemia) and sudden infant death. Later manifestations of SCD include cardiomyopathy and progressive skeletal weakness (Karpati et al. 1975; Stanley et al. 1991).

Shortly after OCTN2 was identified, several groups reported homozygous nonsense or missense mutations of the hOCTN2 gene in patients suffering from SCD (Burwinkel et al. 1999; Koizumi et al. 1999; Lamhonwah et al. 2002; Mayatepek et al. 1999; Nezu et al. 1999; Tang et al. 1999; Vaz et al. 1999; Wang et al. 1999, 2000a, 2000b, 2001). Moreover, it was shown that SCD in *jvs* mice was caused by a mutation in the OCTN2 gene (Hashimoto et al. 1998; Lu et al. 1998; Nezu et al. 1999; Yokogawa et al. 1999). These mice suffer from a carnitine transport defect (Horiuchi et al. 1992, 1994; Kuwajima et al. 1991; Lu et al. 1998) that causes symptoms such as lipid accumulation in the liver [hence the name "juvenile visceral steatosis," or *jvs* (Hayakawa et al. 1990; Koizumi et al. 1988)], cardiac hypertrophy, hypoglycemia, and hyperammonemia (Horiuchi et al. 1993; Tomomura et al. 1992).

To discuss structure–function relationships in OCTN2 together with other transporters of the SLC22 family with known mutants, we compiled the mutations of hOCTN2 together with those of OCT1, OAT1, and OAT3 (Fig. 4). Mutations that cause large decreases of carnitine and/or organic cation transport are lined in yellow. Such mutations occurred in almost all parts of the proteins, including the large extracellular and intracellular loops and TMD 4 (H. Koepsell et al., unpublished data). Since it is usually not known whether the mutants are properly targeted to the plasma membrane, decreased transport rates do not necessarily indicate an involvement of the respective amino acid in transport function. These mutations may specifically disturb targeting or regulation or may induce gross structural changes that alter targeting, regulation, and/or transport. Interestingly, some mutations in the TMDs 4 and 11 of OCTN2 have been shown to specifically affect transport function (Fig. 4, green amino acids). Mutations of tyrosine 211 in the TMD 4 and serine 469 and proline 478 in the TMD 11 lead to an inhibition of carnitine transport without affecting the transport of TEA (Ohashi et al. 2002; Seth et al. 1999). Since it has been shown that mutations in corresponding or nearby positions of rOCT1, fOAT1, and rOAT3 changed the substrate selectivity of these transporters (see Fig. 4, H. Koepsell, unpublished data), it can be concluded that TMDs 4 and 11 contribute to the substrate binding sites of the OCTs, OATs, and OCTNs. In addition, the data support the concept that carnitine and cations bind within the same binding pocket of the OCTNs; this view is also supported by the mutual competitive inhibition of carnitine and TEA transport by TEA and carnitine, respectively (Ohashi et al. 2002; Seth et al. 1999).

The highly conserved intracellular loop between the proposed tenth and 11th TMD (see Fig. 1) is thought to mediate long-range conformational changes during Na^+ carnitine cotransport by OCTN2, and may play an important role in substrate translocation by OCT and OAT transporters. Longo and coworkers observed that the mutation of glutamate 452 to lysine in hOCTN2 decreased Na^+-dependent carnitine cotransport considerably without changing targeting of the protein to the plasma membrane (Wang et al. 2000b, 2000c). Interestingly, the half maximal value for Na^+ activation of carnitine transport was increased to 187 mM as compared to 12 mM in wild type (Wang et al. 2000c), whereas the V_{max} was not changed. Since the Na^+ activation of the Glu452Lys mutant was biphasic the authors hypothesized that Glu452 may be involved in coupling energetically the binding of Na^+ to a conformational change from the outwardly to the inwardly directed conformation of the substrate binding site. Because Glu452 and four additional amino acids in the intracellular loop between the TMD10 and TMD11 are conserved between all members of the SLC22A family, this loop may also play an important role in facilitated diffusion via OCTs.

Overlap between substrate specificities of OCTs and/or OCTNs with other polyspecific transporters

Physiology and biomedical relevance of the organic cation transporters of the OCT and OCTN families has been difficult to assess because there is no simple one-to-one relationship between transporter and substrates: on the one hand, these transporters are promiscuous and transport multiple cation species; on the other hand, their individual contribution may be hard to dissect out in a cellular environment that often contains other transporters with similar substrate selectivity. Transporters that translocate organic cations are found in the SLC22 family, but also in other transporter families.

Transport of cimetidine by OAT3

One example is the organic anion transporter OAT3 which belongs to a subgroup of the SLC22 family, together with the polyspecific organic anion transporters OAT1 through OAT4 and URAT1 (see Table 2 and Fig. 3). OAT1 through OAT4 translocate various organic anions, including endogenous compounds (e.g., *para*-aminohippurate, α-ketoglutarate, cAMP, cGMP, folic acid, prostaglandins, and urate), drugs (e.g., acetylsalicylate, salicylate, indomethacin, mefenamic acid, methrotrexate, estrone sulfate, dehydroepiandrosterone sulfate), and xenobiotics (e.g., ochratoxin; Burckhardt and Wolff 2000; Dresser et al. 2001; Khamdang et al. 2002; Sekine et al. 2000; Sweet and Pritchard 1999b). URAT1 translocates urate and nicotinate and is inhibited by structurally different anions, such as bumetanide, benzbromarone and losartane (Enomoto et al. 2002a). While OAT1, OAT3, and URAT1 may operate as anion/anion exchangers, the transport modes of OAT2 and OAT4 are unclear (Cha et al. 2000, 2001; Enomoto et al. 2002a; Kusuhara et al. 1999; Sekine et al. 1997; Sweet et al. 2003). In humans, OATs 1 through OAT4 and URAT1 are expressed in kidney (Cha et al. 2000, 2001; Hosoyamada et al. 1999; Sun et al. 2001). OAT1 and OAT3 were localized to the basolateral membranes of renal proximal tubules (Cha et al. 2001; Hosoyamada et al. 1999; Jung et al. 2001; Motohashi et al. 2002), whereas URAT1 was found in the brush-border membranes (Enomoto et al. 2002a).

The weak base cimetidine is transported by hOCT2 and hOAT3 (Barendt and Wright 2002; Cha et al. 2001). At physiological pH of 7.4, 75% of cimetidine is uncharged (pK_a of 6.9), and only 25% of it carries a positive charge. At this pH, hOCT2 and hOAT3 exhibit apparent K_m values for cimetidine of 8.6 µM and 57 µM, respectively. Considering that cationic cimetidine is the preferred substrate of hOCT2 (Barendt and Wright 2002), whereas data from flounder OAT1 suggest that OAT1 prefers uncharged cimetidine (Burckhardt et al. 2003), it is reasonable to assume that basolateral uptake of cimetidine in vivo is mediated mainly by hOCT2. An increased contribution of hOAT3 to renal cimetidine uptake would be expected during alkalosis, in the presence of inhibitors of hOCT2, or in genetic disorders that affect hOCT2 function.

Cation transport by members of the SLC21 family

The members of the SLC21 family are polyspecific transporters that may translocate certain organic cations (Hagenbuch and Meier 2003). This family was also named organic anion-transporting polypeptide (OATP) family because it comprises polyspecific transporters that mainly translocate organic anions (for review, see Hagenbuch and Meier 2003). In humans, nine subtypes of this family were cloned, namely PGT (HGNC gene symbol: *SLC21A2*), OATP-A (*SLC21A3;* also called OATP or OATP1), OATP-C (*SLC21A6;* also called LST-1 or OATP2 or OATP6), OATP8 (*SLC21A8;* also called LST-2), OATP-B (*SLCA9;* also called OATP-RP2), OATP-D (*SLC21A11;* also called OATP-RP3), OATP-E (*SLC21A12;* also called OATP-RP1), OATP-F (*SLC21A14;* also called OATP-RP5), and OATP-J (*SLC21A15;* also called OATP-RP4).

The transporters of the SLC21 family contain 12 predicted membrane-spanning α-helices with one large loop between TMD 9 and 10 (Hagenbuch and Meier 2003). The first transporter of this family to be cloned was Oatp1 from rat liver (*Slc21a1*), which was originally described as Na^+-independent uptake system for bromosulphthalein (BSP) and taurocholate (Jacquemin et al. 1994; Shi et al. 1995). Later, it became apparent that Oatp1 is a polyspecific transporter for amphipathic organic compounds including bile salts (Eckhardt et al. 1999; Kullak-Ublick et al. 1994; Reichel et al. 1999), steroid hormones and their conjugates (Eckhardt et al. 1999; Kanai et al. 1996a, 1996b; Reichel et al. 1999), thyroid hormones (Friesema et al. 1999), a number of drugs (Jung et al. 2001), and organic cations such as N-(4.4-azo-*n*-pentyl)-21-deoxy (APD)-ajmalinium, N-methyl-quinine, N-methyl-quinidine, and rocuronium (Bossuyt et al. 1996; van Montfoort et al. 1999). A similarly broad substrate specificity was found for the human transporters OATP-A (*SLC21A3*), OATP-C (*SLC21A3*), and OATP8 (*SLC21A8*) (Hagenbuch and Meier 2003).

As to the human OATP transporters, cation transport could be demonstrated using N-methylquinine for OATP-A, but not for OATP-B, OATP-C, or OATP8 (Kullak-Ublick et al. 2001). OATP-A was cloned from human liver and shows 67% amino acid identity with Oatp1 (Kullak-Ublick et al. 1995). Similar to Oatp1, OATP-A transports N-methylquinine, N-methyl-quinidine, ADP-ajmalinium and rocuronium, with K_m values of 26 µM and 5 µM for N-methylquinine and N-methyl-quinidine, respectively (van Montfoort et al. 1999, 2001). Northern blots demonstrated the expression of OATP-A in brain, kidney, liver, lung, and testis (Kullak-Ublick et al. 1995). In liver, OATP-A protein was localized to the sinusoidal membrane of hepatocytes (Kullak-Ublick et al. 1996, 1997). There, OATP-A works in parallel with other organic ion transporters including OATP-B, OATP-C, and OAT8 (Kullak-Ublick et al. 2001). Although these transporters do not interact with N-methyl-quinindine, it is possible that they accept some other cations as substrates.

Organic cation transport by the ABC transporter MDR1

The human ABC transporter P-glycoprotein or MDR1 (*ABCB1*) is a polyspecific transporter for hydrophobic solutes including organic cations (Gottesman and Pastan 1993; Higgins 1992; Meijer et al. 1997; Sakaeda et al. 2002). MDR1 is infamous for inducing resistance to anticancer drugs in cancer cells, but is found as well in healthy tissues including liver, kidney, small and large intestine, brain, testis, skeletal muscle, placenta, and adrenals. The transporter is located at the luminal membrane of small intestinal and renal proximal epithelial cells, and in the biliary membrane of hepatocytes (Thiebaut et al. 1987, 1989). MDR1 is a primary active transporter which pumps—driven by ATP hydrolysis—compounds of various structure and hydrophobicity out of cells. Water soluble substrates are taken up directly from the cytosol, whereas hydrophobic substrates first partition into the plasma membrane and then enter the transporter from the lipid phase (Higgins and Gottesman 1992).

Both extra- and intracellular compounds may be substrates of MDR1. In tumor cells, brain capillaries, or small intestine, MDR1 protects the cells, the brain, or the whole body, from xenobiotics that passively permeate the plasma membrane. On the other hand, MDR1 mediates the second step in the hepatic and renal excretion of xenobiotics, drugs, and endogenous compounds that have been taken up by transporters in the sinusoidal membrane of hepatocytes and basolateral membrane of renal proximal tubules. The substrate specificity of MDR1 is extremely broad. This fact was first realized after the observation of crossresistance: exposure of cancer cells to a cystostatic compound that is a substrate of MDR1 induces overexpression of the transporter and thus makes it potentially resistant to all other substrates of MDR1 (Gottesman and Pastan 1993).

Transport by MDR1 was studied in cells overexpressing MDR1 such as Caco-2 cells (Tanigawara et al. 1992; Ueda et al. 1992). Substrates may also stimulate ATP hydrolysis by reconstituted MDR1 and may be identified by this property (Ambudkar et al. 1992). Recently, the physiological role of MDR1 for excretion of digoxin could be demonstrated after identification of a mutation that is correlated with a reduced expression of the transporter (Hoffmeyer et al. 2000).

MDR1 translocates and/or binds a large variety of exogenous and endogenous compounds. It transports, for example, anticancer, antiarrhythmic, antihypertensive, anticholinergic, antidepressant and antihistamic drugs, cardiotonic drugs, HIV protease inhibitors, β-adrenergic receptor blockers, Ca^{++}-channel blockers, antimalaria, antifungal, and antihelminthic agents, and steroids (Meijer et al. 1997; Sakaeda et al. 2002). A number of cationic drugs is transported by both MDR1 and hOCT1, or by MDR1 and hOCT2, and may be secreted in vivo by the combined action of the respective transporters. Examples for common substrates of MDR1 and human OCTs are quinidine, verapamil, cimetidine, debrisoquine, acyclovir, and ganciclovir (Table 3; Sakaeda et al. 2002).

Targeted disruption of P-glycoproteins in mice greatly helped to understand the physiological roles of MDR1. While the human P-glycoprotein is encoded by *MDR1*, murine P-glycoproteins are encoded by the genes *mdr1a* and *mdr1b* (Buschman et al. 1992; Croop et al. 1989). For this reason, double knock-out mice with the *mdr1a* and *mdr1b* genes disrupted simultaneously were used to study the function of MDR1 (Smit et al. 1998). In these mice, hepatobiliary secretion of the small cations tributylmethylammonium (TBuMA) and azidoprocainamide metho-iodide (APM) was reduced by 70% compared to control mice (Smit et al. 1998). Since TbuMA and APM are transported substrates of OCT1 from rat and human (van Montfoort et al. 2001) and are probably transported by MDR1, biliary excretion of these compounds may be due to a combined action of OCT1 and MDR1.

Organic cation transport in liver and kidney

Liver

With most transporters for organic cations now being identified and partially characterized, the need to understand how they work in various cells and tissues becomes apparent. Unfortunately, our knowledge concerning cellular and subcellular distribution of human cation transporters, their regulation, and their driving forces is still fragmentary. Using the hepatocyte and the renal proximal epithelial cell as examples, we will discuss how differ-

Fig. 5 Transport systems for organic cations in plasma membranes of human hepatocytes. System A and system B have been functionally identified by transport measurements in membrane vesicles isolated from human hepatocytes. They differ in substrate specificity

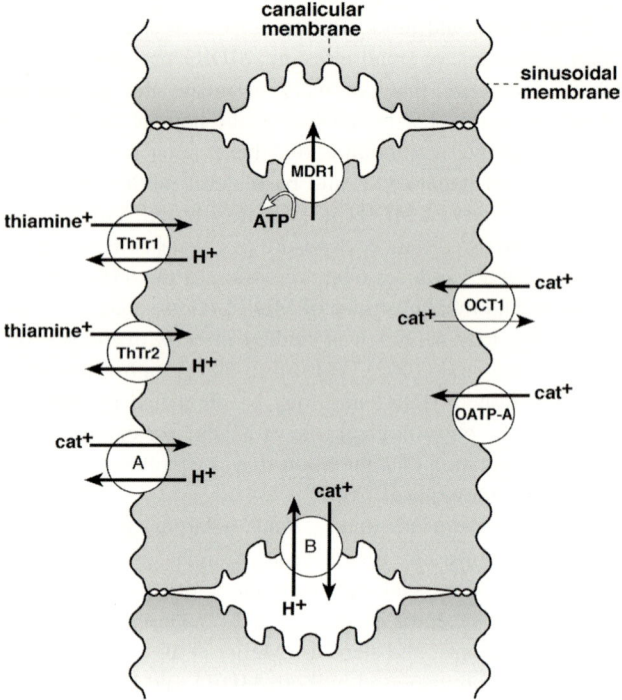

ent organic cation transporters may influence intracellular concentration and transcellular movements of organic cations.

Organic cations absorbed in small intestine directly reach the liver via the portal vein. In contrast, compounds administered intravenously first distribute within the systemic circulation and then enter kidney and liver. Figure 5 shows the cation transport systems of the hepatocyte that have been identified on the molecular (OCT1, OATP-A, MDR1, ThTr1, ThTr2) or functional level (A, B in Fig. 5; Moseley et al. 1992a). The organic cation transporter 3 (OCT3) is highly expressed in human liver, but not indicated in Fig. 5 because its membrane localization is not known.

Uptake of organic cations across the sinusoidal membrane is mainly mediated by the polyspecific transporter OCT1. This transporter accepts small hydrophilic cations but is also able to translocate more bulky hydrophobic cations which may be common substrates with MDR1. At the sinusoidal membrane, some organic cations may be taken up by the organic anion polypeptide OATP-A or by a H^+/organic cation exchanger (A in Fig. 5). In rat, H^+/organic cation exchangers with different substrate specificities have been identified in the sinusoidal and canalicular plasma membrane (Moseley et al. 1990, 1992a). The sinusoidal membrane also contains electroneutral H^+/thiamine antiporters, possibly identical with ThTr1 and ThTr2. Since H^+/thiamine exchange by ThTr1 exhibits a K_m of 0.6 μM for thiamine (Table 1), ThTr1 may be the molecular correlate of the high-affinity thiamine transport that was identified on the functional level in rat hepatocytes (Yoshioka 1984). The second human thiamine transporter ThTr2 (Table 1) may be the molecular correlate of the H^+/thiamine antiport with an apparent K_m of 30 μM that was detected by functional assays in the sinusoidal membrane of rat hepatocytes (Moseley et al. 1992b).

The canalicular membrane of human hepatocytes contains two transporters that translocate organic cations, MDR1 and a H^+/organic cation antiporter that has been functionally characterized in rat (Moseley et al. 1992a).

Taken together, these data suggest that several transport systems in the sinusoidal membrane are involved in the uptake of organic cations. Although OCT1 may be most important for the uptake of small organic cations, another H^+/organic cation antiporter of unknown molecular identity, OCT3, and/or OATP-A may be involved as well. Furthermore, two H^+/thiamine antiporters mediate high affinity uptake of thiamine. Sinusoidal uptake of more hydrophobic and bulky cations may be achieved either via OCT1 or OATP-A. It is possible that additional members of the SLC21 and SLC22 families are involved.

A H^+/organic cation antiporter and MDR1 are responsible for the extrusion of organic cations across the canalicular membrane into the bile. Smaller, more hydrophilic cations are thought to be extruded apically by the H^+/organic cation antiporter, whereas MDR1 may be responsible for the extrusion of more hydrophobic cations. Endogenous cations such as choline and drugs can leave the hepatocytes at the sinusoidal membrane. Since the OCTs can mediate electrogenic cation transport in either direction (see the section entitled "The electrogenic organic cation transporters OCT1, OCT2 and OCT3"), sinusoidal efflux of organic cations may well occur via OCT1. A better understanding of cation efflux via OCTs, including the identification of potential symport or antiport of small ions, is necessary to understand this process.

Kidney

The kidney is the most important organ for the excretion of small organic cations (Koepsell et al. 1999). Small organic cations are filtered freely in the glomeruli. In the proximal tubule, either partial reabsorption or further secretion may follow. Ultrafiltration occurs readily for hydrophilic cations, but is not efficient for cations that are bound to plasma proteins. Many endogenous organic cations, cationic drugs and xenobiotics belong to this latter group. For some cations bidirectional transport in the nephron has been demonstrated. For example, in rat kidney choline is secreted at high plasma concentrations, whereas it is reabsorbed at normal plasma concentrations below 25 µM (Acara and Rennick 1973; Besseghir et al. 1981). In humans, the first step in tubular cation secretion, i.e., organic cation uptake across the basolateral membrane, is virtually exclusively mediated by OCT2 (Fig. 6). A splice variant of hOCT2, hOCT2A, showed low activity in vitro (Urakami et al. 2002); nonetheless, hOCT2A may contribute to basolateral uptake if it is present in that membrane. OAT3 transports mainly cimetidine (see the section entitled "Overlap between substrate specificities of OCTs and/or OCTNs with other polyspecific transporters"), but also several weak bases and permanent organic cations; it may thus be another contributor to basolateral organic cation uptake.

The second step in organic cation secretion, i.e., extrusion across the brush-border membrane into the lumen of the renal tubule, may be mediated by four different transport systems: two H^+/organic cation antiporters, the Na^+/carnitine cotransporter OCTN2, and MDR1. The H^+/organic cation antiporters include OCTN1, which does not interact with NMN (Yabuuchi et al. 1999), and a H^+/organic cation antiporter of unknown molecular identity which does translocate NMN (Ott et al. 1991). Cation efflux across the brush-border may be also mediated by the Na^+-carnitine cotransporter hOCTN2 which has been shown to translocate organic cations into cells. hOCTN2 is probably symmetric with re-

Fig. 6 Transport systems for organic cations in plasma membranes of human renal proximal tubules. Systems *B* and *C* have not been demonstrated in human proximal tubules. Their existence is hypothesized since these systems have been functionally demonstrated in proximal tubules from rat and rabbit. System B may be identical to system B in Fig. 5

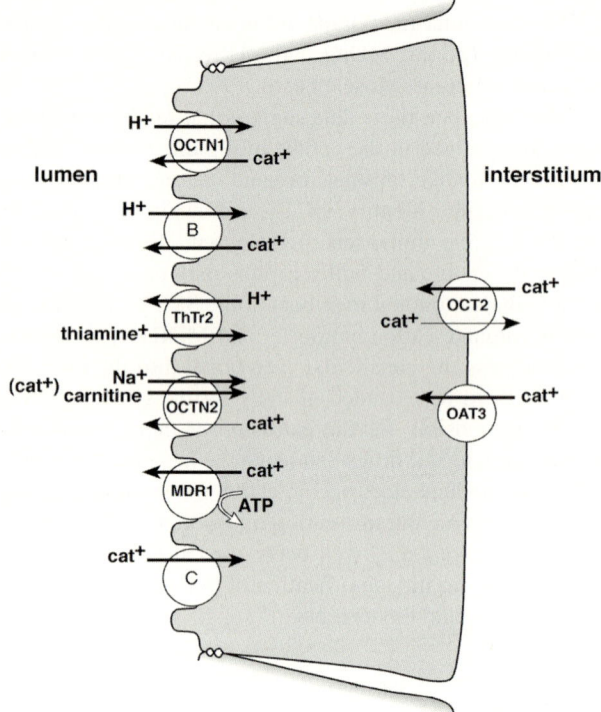

spect to organic cation transport, and organic cation efflux via murine OCTN2 is *trans*-stimulated in the presence of Na^+ (Ohashi et al. 2001). MDR1 in the brush-border membrane may be responsible for efflux of bulky hydrophobic organic cations.

The first step in the reabsorption of small hydrophilic organic cations—cation uptake across the luminal membrane—is probably mediated by an electrogenic and polyspecific cation transport system (Fig. 6, system C) that has been characterized in rat and was primarily regarded as a choline transporter (Takano et al. 1993; Ullrich and Rumrich 1996; Wright et al. 1992). The second step in cation reabsorption—the efflux across the basolateral membrane—may be mediated by OCT2. This transporter is able to translocate cations in both directions, and the opposing effects of the inside-negative membrane potential may be neutralized if the transporter can operate in an electroneutral cation/cation exchange mode; this mode would imply the presence of some other organic cation substrate on the extracellular side.

Conclusions

After identifying most of the plasma membrane transporters involved in absorption and excretion of organic cations, the present challenge is to achieve a comprehensive understanding of cation transport in human by cloning the last transporters missing, by filling the multiple white spots on our map with respect to their tissue distribution and membrane localization, and by clarifying their molecular transport mechanisms. Since substrate and inhibitor specificities of many cloned transporters are known in some detail, specific inhibitors are available that allow analysis of individual transporters in vivo. Such studies are required to define the contribution of individual transporters to the absorption, secretion, and tissue distribution of specific drugs in a complex biological setting where multiple, different cation transporters of overlapping substrate specificity are often found within the same membrane. The importance of the "team work" between basolateral and luminal transporters in renal proximal tubules or hepatocytes is obvious; in order to study such integrated cellular processes in further detail, it may be promising to generate polarized eukaryotic cell lines that express the various combinations of transporters in the appropriate subcellular sites. Monolayers of such cells would be suitable to test drugs (or candidate compounds) for their putative renal and hepatic excretion, and to screen for drug–drug interactions at the transporters.

The identification of single nucleotide polymorphisms (SNPs) in human organic cation transporters that result in decreased expression or impaired function could make possible the timely identification of persons that are at risk for certain adverse drug reactions, particularly so when the list of known and characterized polymorphisms in organic cation transporters has become more complete. In this context, the potential identification of genotypes associated with spontaneous (i.e., not induced by an exogenous compound) functional changes or diseases may help to define the physiological role of these transporters.

The structural underpinnings of the unique functional properties exhibited by the transporters of the SLC22 family remain a challenge and promise insights of considerable theoretical relevance. Firstly, what is the structural basis for substrate recognition in such polyspecific transporters? Secondly, are there leak permeabilities, e.g., for small inorganic cations or anions, or may small inorganic anions or cations be cotransported together with the organic substrates? Thirdly, which are the structural features that turn some SLC22 transporters into an anion/anion exchanger (e.g., the OAT transporters), as opposed to facilitated diffusion (e.g., the OCT transporters)? Fourthly, which are the structural features that turn other SLC22 transporters into a Na^+-driven cotransporter (e.g., OCTN2)? These questions can be addressed by further site-directed mutagenesis studies in combination with detailed functional analysis of the mutants. Herein, much information is lost (or, because of the uncontrolled effects of membrane potential, sometimes even distorted) when function is studied mainly by isotope uptake assays; rather, the full repertoire of biochemical and biophysical techniques should be applied to these transporters. Membrane topology of the transporters (currently mainly based on predictions from the primary amino acid sequence) should be determined empirically, as well as the question of whether these transporters exist as functional or structural oligomers. Ultimately, 2D and 3D crystals and high-resolution tertiary structures could make invaluable contributions to a deeper understanding of organic cation transporters.

References

Acara M, Rennick B (1973) Regulation of plasma choline by the renal tubule: bidirectional transport of choline. Am J Physiol 225:1123–1128

Akaike N, Yatani A, Nishi K, Oyama Y, Kuraoka S (1984) Permeability to various cations of the voltage-dependent sodium channel of rat single heart cells. J Pharmacol Exp Ther 228:225–229

Ambudkar SV, Lelong IH, Zhang J, Cardarelli CO, Gottesman MM, Pastan I (1992) Partial purification and reconstitution of the human multidrug-resistance pump: characterization of the drug-stimulatable ATP hydrolysis. Proc Natl Acad Sci USA 89:8472–8476

Apparsundaram S, Ferguson SM, George AL Jr, Blakely RD (2000) Molecular cloning of a human, hemi-
    cholinium-3-sensitive choline transporter. Biochem Biophys Res Commun 276:862–867
Apparsundaram S, Ferguson SM, Blakely RD (2001) Molecular cloning and characterization of a murine
    hemicholinium-3-sensitive choline transporter. Biochem Soc Trans 29:711–716
Arndt P, Volk C, Gorboulev V, Budiman T, Popp C, Ulzheimer-Teuber I, Akhoundova A, Koppatz S, Bam-
    berg E, Nagel G, Koepsell H (2001) Interaction of cations, anions, and weak base quinine with rat re-
    nal cation transporter rOCT2 compared with rOCT1. Am J Physiol Renal Physiol 281:F454–F468
Bahn A, Prawitt D, Buttler D, Reid G, Enklaar T, Wolff NA, Ebbinghaus C, Hillemann A, Schulten H-J,
    Gunawan B, Füzesi L, Zabel B, Burckhardt G (2000) Genomic structure and in vivo expression of the
    human organic anion transporter 1 (hOAT1) gene. Biochem Biophys Res Commun 275:623–630
Barendt WM, Wright SH (2002) The human organic cation transporter (hOCT2) recognizes the degree of
    substrate ionization. J Biol Chem 277:22491–22496
Bengel D, Murphy DL, Andrews AM, Wichems CH, Feltner D, Heils A, Mössner R, Westphal H, Lesch K-
    P (1998) Altered brain serotonin homeostasis and locomotor insensitivity to 3,4-methylene-
    dioxymethamphetamine ("Ecstasy") in serotonin transporter-deficient mice. Mol Pharmacol 53:649–
    655
Berardi S, Stieger B, Hagenbuch B, Carafoli E, Krähenbühl S (2000) Characterization of L-carnitine trans-
    port into rat skeletal muscle plasma membrane vesicles. Eur J Biochem 267:1985–1994
Besseghir K, Pearce LB, Rennick B (1981) Renal tubular transport and metabolism of organic cations by
    the rabbit. Am J Physiol Renal Physiol 241:F308–F314
Blakely RD, De Felice LJ, Hartzell HC (1994) Molecular physiology of norepinephrine and serotonin trans-
    porters. J Exp Biol 196:263–281
Borowsky B, Hoffman BJ (1995) Neurotransmitter transporters: molecular biology, function, and regula-
    tion. Int Rev Neurobiol 38:139–199
Borst SE, Snellen H G (2001) Metformin, but not exercise training, increases insulin responsiveness in skel-
    etal muscle of Sprague-Dawley rats. Life Sci 69:1497–1507
Bossuyt X, Müller M, Hagenbuch B, Meier PJ (1996) Polyspecific steroid and drug clearance by an organic
    anion transporter of mammalian liver. J Pharmacol Exp Ther 276:891–896
Bowman HM, Hook JB (1972) Sex differences in organic ion transport by rat kidney. Proc Soc Exp Biol
    Med 141:258–262
Bravo D, Parsons S M (2002) Microscopic kinetics and structure–function analysis in the vesicular acetyl-
    choline transporter. Neurochem Int 41:285–289
Briz O, Serrano MA, Rebollo N, Hagenbuch B, Meier PJ, Koepsell H, Marin JJG (2002) Carriers involved
    in targeting the cytostatic bile acid-cisplatin derivatives cis-diammine-chloro-cholylglycinate-plati-
    num(II) and cis- diammine-bisursodeoxycholate-platinum(II) toward liver cells. Mol Pharmacol
    61:853–860
Brooks H, Krähenbühl S (2001) Identification and tissue distribution of two differentially spliced variants
    of the rat carnitine transporter OCTN2. FEBS Lett 508:175–180
Bruss M, Kunz J, Lingen B, Bönisch H (1993) Chromosomal mapping of the human gene for the tricyclic
    antidepressant-sensitive noradrenaline transporter. Hum Genet 91:278–280
Bryan-Lluka LJ, Westwood NN, O'Donnell SR (1992) Vascular uptake of catecholamines in perfused lungs
    of the rat occurs by the same process as uptake1 in noradrenergic neurones. Naunyn Schmiedebergs
    Arch Pharmacol 345:319–326
Budiman T, Bamberg E, Koepsell H, Nagel G (2000) Mechanism of electrogenic cation transport by the
    cloned organic cation transporter 2 from rat. J Biol Chem 275:29413–29420
Burckhardt BC, Brai S, Wallis S, Krick W, Wolff NA, Burckhardt G (2003) Transport of cimetidine by
    flounder and human renal organic anion transporter 1. Am J Physiol Renal Physiol 284:F503–F509
Burckhardt G, Wolff NA (2000) Structure of renal organic anion and cation transporters. Am J Physiol Re-
    nal Physiol 278:F853–F866
Burwinkel B, Kreuder J, Schweitzer S, Vorgerd M, Gempel K, Gerbitz K-D, Kilimann MW (1999) Carni-
    tine transporter OCTN2 mutations in systemic primary carnitine deficiency: a novel Arg169Gln muta-
    tion and a recurrent Arg282ter mutation associated with an unconventional splicing abnormality. Bio-
    chem Biophys Res Commun 261:484–487
Busch AE, Quester S, Ulzheimer JC, Waldegger S, Gorboulev V, Arndt P, Lang F, Koepsell H (1996a)
    Electrogenic properties and substrate specificity of the polyspecific rat cation transporter rOCT1. J
    Biol Chem 271:32599–32604
Busch AE, Quester S, Ulzheimer J C, Gorboulev V, Akhoundova A, Waldegger S, Lang F, Koepsell H
    (1996b) Monoamine neurotransmitter transport mediated by the polyspecific cation transporter rOCT1.
    FEBS Lett 395:153–156

Busch AE, Karbach U, Miska D, Gorboulev V, Akhoundova A, Volk C, Arndt P, Ulzheimer J C, Sonders M S, Baumann C, Waldegger S, Lang F, Koepsell H (1998) Human neurons express the polyspecific cation transporter hOCT2, which translocates monoamine neurotransmitters, amantadine, and memantine. Mol Pharmacol 54:342–352

Buschman E, Arceci RJ, Croop J M, Che M, Arias IM, Housman DE, Gros P (1992) *Mdr2* encodes P-glycoprotein expressed in the bile canalicular membrane as determined by isoform-specific antibodies. J Biol Chem 267:18093–18099

Bzoskie L, Blount L, Kashiwai K, Tseng YT, Hay WW, Jr., Padbury JF (1995) Placental norepinephrine clearance: in vivo measurement and physiological role. Am J Physiol 269:E145–E149

Bzoskie L, Blount L, Kashiwai K, Humme J, Padbury JF (1997) The contribution of transporter-dependent uptake to fetal catecholamine clearance. Biol Neonate 71:102–110

Calabresi P, Centonze D, Bernardi G (2000) Electrophysiology of dopamine in normal and denervated striatal neurons. Trends Neurosci 23:S57-S63

Cargill M, Altshuler D, Ireland J, Sklar P, Ardlie K, Patil N, Lane CR, Lim EP, Kalyanaraman N, Nemesh J, Ziaugra L, Friedland L, Rolfe A, Warrington J, Lipshutz R, Daley GQ, Lander ES (1999) Characterization of single-nucleotide polymorphisms in coding regions of human genes. Nat Genet 22:231–238

Caspary WF, Creutzfeldt W (1971) Analysis of the inhibitory effect of biguanides on glucose absorption: inhibition of active sugar transport. Diabetologia 7:379–385

Catravas JD, Gillis CN (1980) Pulmonary clearance of [^{14}C]-5-hydroxytryptamine and [^{3}H]norepinephrine in vivo: effects of pretreatment with imipramine or cocaine. J Pharmacol Exp Ther 213:120–127

Cetinkaya I, Ciarimboli G, Yalcinkaya, G, Mehrens T, Velic A, Hirsch JR, Gorboulev V, Koepsell H, Schlatter E (2003) The human organic cation transporter hOCT2 is regulated by Ca^{2+}/calmodulin-, cAMP- and phosphatidylinositol-3-dependent kinases. Am J Physiol Renal Physiol 284: F293–F302

Cha SH, Sekine T, Kusuhara H, Yu E, Kim JY, Kim DK, Sugiyama Y, Kanai Y, Endou H (2000) Molecular cloning and characterization of multispecific organic anion transporter 4 expressed in the placenta. J Biol Chem 275:4507–4512

Cha SH, Sekine T, Fukushima J-I, Kanai Y, Kobayashi Y, Goya T, Endou H (2001) Identification and characterization of human organic anion transporter 3 expressing predominantly in the kidney. Mol Pharmacol 59:1277–1286

Chen JJ, Li Z, Pan H, Murphy DL, Tamir H, Koepsell H, Gershon MD (2001) Maintenance of serotonin in the intestinal mucosa and ganglia of mice that lack the high-affinity serotonin transporter: abnormal intestinal motility and the expression of cation transporters. J Neurosci 21:6348–6361

Chen J-X, Pan H, Rothman TP, Wade PR, Gershon MD (1998) Guinea pig 5-HT transporter: cloning, expression, distribution, and function in intestinal sensory reception. Am J Physiol Gastroenterol Liver Physiol 275:G433–G448

Chen R, Nelson JA (2000) Role of organic cation transporters in the renal secretion of nucleosides. Biochem Pharmacol 60:215–219

Chen R, Jonker JW, Nelson J A (2002) Renal organic cation and nucleoside transport. Biochem Pharmacol 64:185–190

Croop JM, Raymond M, Haber D, Devault A, Arceci RJ, Gros P, Housman DE (1989) The three mouse multidrug resistance (*mdr*) genes are expressed in a tissue-specific manner in normal mouse tissues. Mol Cell Biol 9:1346–1350

Dai G, Levy O, Carrasco N (1996) Cloning and characterization of the thyroid iodide transporter. Nature 379:458–460

Davidson MB, Peters AL (1997) An overview of metformin in the treatment of type 2 diabetes mellitus. Am J Med 102:99–110

Diaz GA, Banikazemi M, Oishi K, Desnick R J, Gelb BD (1999) Mutations in a new gene encoding a thiamine transporter cause thiamine-responsive megaloblastic anaemia syndrome. Nat Genet 22:309–312

Dresser MJ, Gray AT, Giacomini KM (2000) Kinetic and selectivity differences between rodent, rabbit, and human organic cation transporters (OCT1). J Pharmacol Exp Ther 292:1146–1152

Dresser MJ, Leabman MK, Giacomini KM (2001) Transporters involved in the elimination of drugs in the kidney: organic anion transporters and organic cation transporters. J Pharm Sci 90:397–421

Dresser MJ, Xiao G, Leabman MK, Gray AT, Giacomini KM (2002) Interactions of *n*-tetraalkylammonium compounds and biguanides with a human renal organic cation transporter (hOCT2). Pharmac Res 19:1244–1247

Durán JM, Peral MJ, Calonge ML, Ilundáin AA (2002) Functional characterization of intestinal L-carnitine transport. J Membr Biol 185:65–74

Dutta B, Huang W, Molero M, Kekuda R, Leibach FH, Devoe LD, Ganapathy V, Prasad PD (1999) Cloning of the human thiamine transporter, a member of the folate transporter family. J Biol Chem 274:31925–31929

Eckhardt U, Schroeder A, Stieger B, Höchli M, Landmann L, Tynes R, Meier PJ, Hagenbuch B (1999) Polyspecific substrate uptake by the hepatic organic anion transporter Oatp1 in stably transfected CHO cells. Am J Physiol Gastrointest Liver Physiol 276:G1037–G1042

Eiden LE (1998) The cholinergic gene locus. J Neurochem 70:2227–2240

Eisenhofer G (2001) The role of neuronal and extraneuronal plasma membrane transporters in the inactivation of peripheral catecholamines. Pharmacol Ther 91:35–62

Eisenhofer G, Åneman A, Friberg P, Hooper D, Fåndriks L, Lonroth H, Hunyady B, Mezey, E (1997) Substantial production of dopamine in the human gastrointestinal tract. J Clin Endocrinol Metab 82:3864–3871

El-Mir M-Y, Nogueira V, Fontaine E, Averet N, Rigoulet M, Leverve X (2000) Dimethylbiguanide inhibits cell respiration via an indirect effect targeted on the respiratory chain complex I. J Biol Chem 275:223–228

Elferink RPJO, Meijer DKF, Kuipers F, Jansen PLM, Groen AK, Groothuis GMM (1995) Hepatobiliary secretion of organic compounds; molecular mechanisms of membrane transport. Biochim Biophys Acta 1241:215–268

Enomoto A, Kimura H, Chairoungdua A, Shigeta Y, Jutabha P, Cha SH, Hosoyamada M, Takeda M, Sekine T, Igarashi T, Matsuo H, Kikuchi Y, Oda T, Ichida K, Hosoya T, Shimokata K, Niwa T, Kanai Y, Endou H (2002a) Molecular identification of a renal urate-anion exchanger that regulates blood urate levels. Nature 417:447–452

Enomoto A, Wempe MF, Tsuchida H, Shin HJ, Cha SH, Anzai N, Goto A, Sakamoto A, Niwa T, Kanai Y, Anders MW, Endou H (2002b) Molecular identification of a novel carnitine transporter specific to human testis: insights into the mechanism of carnitine recognition. J Biol Chem 277:36262–36271

Eraly SA, Nigam SK (2002) Novel human cDNAs homologous to drosophila Orct and mammalian carnitine transporters. Biochem Biophys Res Commun 297:1159–1166

Eraly SA, Hamilton BA, Nigam SK (2003) Organic anion and cation transporters occur in pairs of similar and similarly expressed genes Biochem Biophys Res Commun 300:333–342

Erickson JD, Varoqui H, Schafer MK, Modi W, Diebler MF, Weihe E, Rand J, Eiden LE, Bonner TI, Usdin TB (1994) Functional identification of a vesicular acetylcholine transporter and its expression from a "cholinergic" gene locus. J Biol Chem 269:21929–21932

Eudy JD, Spiegelstein O, Barber RC, Wlodarczyk BJ, Talbot J, Finnell RH (2000) Identification and characterization of the human and mouse *SLC19A3* gene: a novel member of the reduced folate family of micronutrient transporter genes. Mol Genet Metab 71:581–590

Felsenstein J (1998) Phylogeny inference package (PHYLIP) version 3.5 C, distributed by author, Dept. of Genetics, University of Washington, Seattle. *PHYLIP*.

Feng B, Dresser MJ, Shu Y, Johns SJ, Giacomini KM (2001) Arginine 454 and lysine 370 are essential for the anion specificity of the organic anion transporter, rOAT3. Biochemistry 40:5511–5520

Fleming JC, Tartaglini E, Steinkamp MP, Schorderet DF, Cohen N and Neufeld EJ (1999) The gene mutated in thiamine-responsive anaemia with diabetes and deafness (TRMA) encodes a functional thiamine transporter. Nat Genet 22:305–308

Fleming JC, Steinkamp M P, Kawatsuji R, Tartaglini E, Pinkus JL, Pinkus GS, Fleming MD, Neufeld EJ (2001) Characterization of a murine high-affinity thiamine transporter, *Slc19a2*. Mol Genet Metab 74:273–280

Francis PT, Palmer AM, Snape M, Wilcock GK (1999) The cholinergic hypothesis of Alzheimer's disease: a review of progress. J Neurol Neurosurg Psychiatry 66:137–147

Friesema ECH, Docter R, Moerings EPCM, Stieger B, Hagenbuch B, Meier PJ, Krenning EP, Hennemann G, Visser TJ (1999) Identification of thyroid hormone transporters. Biochem Biophys Res Commun 254:497–501

Gainetdinov RR, Jones SR, Caron MG (1999) Functional hyperdopaminergia in dopamine transporter knock-out mice. Biol Psychiatry 46:303–311

Ganapathy ME, Huang W, Rajan DP, Carter AL, Sugawara M, Iseki K, Leibach FH, Ganapathy V (2000) β-lactam antibiotics as substrates for OCTN2, an organic cation/carnitine transporter. J Biol Chem 275:1699–1707

Gelernter J, Kruger S, Pakstis A J, Pacholczyk T, Sparkes RS, Kidd KK, Amara S (1993) Assignment of the norepinephrine transporter protein (NET1) locus to chromosome 16. Genomics 18:690–692

Gerk PM, Oo CY, Paxton EW, Moscow JA, McNamara PJ (2001) Interactions between cimetidine, nitrofurantoin, and probenecid active transport into rat milk. J Pharmacol Exp Ther 296:175–180

Giros B, El Mestikawy S, Bertrand L, Caron MG (1991) Cloning and functional characterization of a cocaine-sensitive dopamine transporter. FEBS Lett 295:149–154

Giros B, El Mestikawy S, Godinot N, Zheng K, Han H, Yang-Feng T, Caron MG (1992) Cloning, pharmacological characterization, and chromosome assignment of the human dopamine transporter. Mol Pharmacol 42:383–390

Goldstein DS, Mezey E, Yamamoto T, Aneman A, Friberg P, Eisenhofer G (1995) Is there a third peripheral catecholaminergic system? Endogenous dopamine as an autocrine/paracrine substance derived from plasma DOPA and inactivated by conjugation. Hypertens Res 18 Suppl 1:S93–S99

Gong S, Lu X, Xu Y, Swiderski CF, Jordan CT, Moscow JA (2002) Identification of OCT6 as a novel organic cation transporter preferentially expressed in hematopoietic cells and leukemias. Exp Hematol 30:1162–1169

Goralski KB, Lou G, Prowse M T, Gorboulev V, Volk C, Koepsell H, Sitar DS (2002) The cation transporters rOCT1 and rOCT2 interact with bicarbonate but play only a minor role for amantadine uptake into rat renal proximal tubules. J Pharmacol Exp Ther 303:959–968

Gorboulev V, Ulzheimer JC, Akhoundova A, Ulzheimer-Teuber I, Karbach U, Quester S, Baumann C, Lang F, Busch AE, Koepsell H (1997) Cloning and characterization of two human polyspecific organic cation transporters. DNA Cell Biol 16:871–881

Gorboulev V, Volk C, Arndt P, Akhoundova A, Koepsell H (1999) Selectivity of the polyspecific cation transporter rOCT1 is changed by mutation of aspartate 475 to glutamate. Mol Pharmacol 56:1254–1261

Gottesman MM, Pastan I (1993) Biochemistry of multidrug resistance mediated by the multidrug transporter. Annu Rev Biochem 62:385–427

Graefe K-H, Bönisch H (1988) The transport of amines across the axonal membranes of noradrenergic and dopaminergic neurones. In Trendelenburg U and Weiner N (eds) Handbook of experimental pharmacology. Catecholamines I. Springer-Verlag, Berlin, pp 193–245

Grohmann M, Trendelenburg U (1984) The substrate specificity of uptake2 in the rat heart. Naunyn Schmiedebergs Arch Pharmacol 328:164–173

Grover B, Auberger C, Sarangarajan R, Cacini W (2002) Functional impairment of renal organic cation transport in experimental diabetes. Pharmacol Toxicol 90:181–186

Gründemann D, Schömig E (2000) Gene structures of the human nonneuronal monoamine transporters EMT and OCT2. Hum Genet 106:627–635

Gründemann D, Gorboulev V, Gambaryan S, Veyhl M, Koepsell H (1994) Drug excretion mediated by a new prototype of polyspecific transporter. Nature 372:549–552

Gründemann D, Babin-Ebell J, Martel F, Örding N, Schmidt A, Schömig E (1997) Primary structure and functional expression of the apical organic cation transporter from kidney epithelial LLC-PK$_1$ cells. J Biol Chem 272:10408–10413

Gründemann D, Schechinger B, Rappold GA, Schömig E (1998a) Molecular identification of the corticosterone-sensitive extraneuronal catecholamine transporter. Nature neurosci 1:349–352

Gründemann D, Köster S, Kiefer N, Breidert T, Engelhardt M, Spitzenberger F, Obermüller N, Schömig E (1998b) Transport of monoamine transmitters by the organic cation transporter type 2, OCT2. J Biol Chem 273:30915–30920

Gründemann D, Liebich G, Kiefer N, Köster S, Schömig E (1999) Selective substrates for nonneuronal monoamine transporters. Mol Pharmacol 56:1–10

Haberberger RV, Pfeil U, Lips KS, Kummer W (2002). Expression of the high-affinity choline transporter, CHT1, in the neuronal and nonneuronal cholinergic system of human and rat skin. J Invest Dermatol 119:1–6

Haga T (1971) Synthesis and release of (^{14}C)acetylcholine in synaptosomes. J Neurochem 18:781–798

Haga T, Noda H (1973) Choline uptake systems of rat brain synaptosomes. Biochim Biophys Acta 291:564–575

Hagenbuch B, Meier PJ (2003) The superfamily of organic anion transporting polypeptides. Biochim Biophys Acta 1609:1–18

Halushka MK, Fan J-B, Bentley K, Hsie L, Shen N, Weder A, Cooper R, Lipshutz R, Chakravarti A (1999) Patterns of single-nucleotide polymorphisms in candidate genes for blood-pressure homeostasis. Nat Genet 22:239–247

Harik SI, Hritz MA (1993) Effect of acetyl-L-carnitine on 1-methyl-4-phenyl-1,2,3,6- tetrahydropyridine (MPTP) neurotoxicity. Biochem Pharmacol 45:2170–2172

Hashimoto N, Suzuki F, Tamai I, Nikaido H, Kuwajima M, Hayakawa J-I, Tsuji A (1998) Gene-dose effect on carnitine transport activity in embryonic fibroblasts of JVS mice as a model of human carnitine transporter deficiency. Biochem Pharmacol 55:1729–1732

Hayakawa J, Koizumi T, Nikaido H (1990) Inheritance of juvenile visceral steatosis found in C3H-H-2° mice. Mouse Genome 86:261

Hayer M, Bönisch H, Brüss M (1999) Molecular cloning, functional characterization and genomic organization of four alternatively spliced isoforms of the human organic cation transporter 1 (hOCT1/SLC22A1). Ann Hum Genet 63:473–482

Hayer-Zillgen M, Brüss M, Bönisch H (2002) Expression and pharmacological profile of the human organic cation transporters hOCT1, hOCT2 and hOCT3. Br J Pharmacol 136:829–836

Hediger MA, Coady MJ, Ikeda TS, Wright EM (1987) Expression cloning and cDNA sequencing of the Na+/glucose cotransporter. Nature 330:379–381

Higgins CF (1992) ABC transporters: from microorganisms to man. Annu Rev Cell Biol 8:67–113

Higgins CF, Gottesman MM (1992) Is the multidrug transporter a flippase? Trends Biochem Sci 17:18–21

Hoffman BJ, Hansson SR, Mezey E, Palkovits M (1998) Localization and dynamic regulation of biogenic amine transporters in the mammalian central nervous system. Front Neuroendocrinol 19:187–231

Hoffmeyer S, Burk O, von Richter O, Arnold HP, Brockmöller J, Johne A, Cascorbi I, Gerloff T, Roots I, Eichelbaum M, Brinkmann U (2000) Functional polymorphisms of the human multidrug-resistance gene: multiple sequence variations and correlation of one allele with P-glycoprotein expression and activity in vivo. Proc Natl Acad Sci USA 97:3473–3478

Hohage H, Mörth DM, Querl IU, Greven J (1994) Regulation by protein kinase C of the contraluminal transport system for organic cations in rabbit kidney S2 proximal tubules. J Pharmacol Exp Ther 268:897–901

Horiuchi M, Yoshida H, Kobayashi K, Kuriwaki K, Yoshimine K, Tomomura M, Koizumi T, Nikaido H, Hayakawa J, Kuwajima M, Saheki T (1993) Cardiac hypertrophy in juvenile visceral steatosis (jvs) mice with systemic carnitine deficiency. FEBS Lett 326:267–271

Horiuchi M, Kobayashi K, Tomomura M, Kuwajima M, Imamura Y, Koizumi T, Nikaido H, Hayakawa J, Saheki T (1992) Carnitine administration to juvenile visceral steatosis mice corrects the suppressed expression of urea cycle enzymes by normalizing their transcription. J Biol Chem 267:5032–5035

Horiuchi M, Kobayashi K, Yamaguchi S, Shimizu N, Koizumi T, Nikaido H, Hayakawa J-I, Kuwajima M, Saheki T (1994) Primary defect of juvenile visceral steatosis (jvs) mouse with systemic carnitine deficiency is probably in renal carnitine transport system. Biochim Biophys Acta 1226:25–30

Hosoyamada M, Sekine T, Kanai Y, Endou H (1999) Molecular cloning and functional expression of a multispecific organic anion transporter from human kidney. Am J Physiol Renal Physiol 276:F122–F128

Hundal RS, Krssak M, Dufour S, Laurent D, Lebon V, Chandramouli V, Inzucchi SE, Schumann WC, Petersen KF, Landau BR, Shulman GI (2000) Mechanism by which metformin reduces glucose production in type 2 diabetes. Diabetes 49:2063–2069

Hyde TM, Crook JM (2001) Cholinergic systems and schizophrenia: primary pathology or epiphenomena? J Chem Neuroanat 22:53–63

Ikeda E, Shiba K, Mori H, Ichikawa A, Sumiya H, Kuji I, Tonami N (2000) Reduction of vesicular acetylcholine transporter in beta-amyloid protein-infused rats with memory impairment. Nucl Med Commun 21:933–937

Inui K-I, Takano M, Okano T, Hori R (1985) H+gradient-dependent transport of aminocephalosporins in rat renal brush border membrane vesicles: role of H+/organic cation antiport system. J Pharmacol Exp Ther 233:181–185

Isacson O, Seo H, Lin L, Albeck D, Granholm AC (2002) Alzheimer's disease and Down's syndrome: roles of APP, trophic factors and ACh. Trends Neurosci 25:79–84

Iversen LL (1965) The uptake of catechol amines at high perfusion concentration in the rat isolated heart: a novel catechol amine uptake process. Br J Pharmacol 25:18–33

Iversen LL, Salt P-J (1970) Inhibition of catecholamine uptake2 by steroids in the isolated rat heart. Br J Pharmacol 40:528–530

Jacquemin E, Hagenbuch B, Stieger B, Wolkoff AW, Meier PJ (1994) Expression cloning of a rat liver Na+-independent organic anion transporter. Proc Natl Acad Sci USA 91:133–137

Ji L, Masuda S, Saito H, Inui K (2002) Downregulation of rat organic cation transporter rOCT2 by 5/6 nephrectomy. Kidney Int 62:514–524

Jonker JW, Wagenaar E, Mol CAAM, Buitelaar M, Koepsell H, Smit JW, Schinkel AH (2001) Reduced hepatic uptake and intestinal excretion of organic cations in mice with a targeted disruption of the organic cation transporter 1 (Oct1[Slc22a1]) gene. Mol Cell Biol 21:5471–5477

Jung JS, Kim YK, Lee SH (1989) Characteristics of tetraethylamonium transport in rabbit renal plasma-membrane vesicles. Biochem J 259:377–383

Jung KY, Takeda M, Kim DK, Tojo A, Narikawa S, Yoo BS, Hosoyamada M, Cha SH, Sekine T, Endou H (2001) Characterization of ochratoxin A transport by human organic anion transporters. Life Sci 69:2123–2135

Kakehi M, Koyabu N, Nakamura T, Uchiumi T, Kuwano M, Ohtani H, Sawada Y (2002) Functional characterization of mouse cation transporter mOCT2 compared with mOCT1. Biochem Biophys Res Commun 296:644–650

Kanai N, Lu R, Bao Y, Wolkoff AW, Schuster VL (1996a) Transient expression of Oatp organic anion transporter in mammalian cells: identification of candidate substrates. Am J Physiol Renal Physiol 270:F319–F325

Kanai N, Lu R, Bao Y, Wolkoff AW, Vore M, Schuster VL (1996b) Estradiol 17 β-D-glucuronide is a high-affinity substrate for Oatp organic anion transporter. Am J Physiol Renal Phsiol 270:F326–F331

Kanner BI (1994) Sodium-coupled neurotransmitter transport: structure, function and regulation. J Exp Biol 196:237–249

Karbach U, Kricke J, Meyer-Wentrup F, Gorboulev V, Volk C, Loffing-Cueni D, Kaissling B, Bachmann S, Koepsell H (2000) Localization of organic cation transporters OCT1 and OCT2 in rat kidney. Am J Physiol Renal Physiol 279:F679–F687

Karpati G, Carpenter S, Engel AG, Watters G, Allen J, Rothman S, Klassen G, Mamer OA (1975) The syndrome of systemic carnitine deficiency. Clinical, morphologic, biochemical, and pathophysiologic features. Neurology 25:16–24

Kekuda R, Prasad PD, Wu X, Wang H, Fei Y-J, Leibach FH, Ganapathy V (1998) Cloning and functional characterization of a potential-sensitive, polyspecific organic cation transporter (OCT3) most abundantly expressed in placenta. J Biol Chem 273:15971–15979

Kerb R, Brinkmann U, Chatskaia N, Gorbunov D, Gorboulev V, Mornhinweg E, Keil A, Eichelbaum M, Koepsell H (2002) Identification of genetic variations of the human organic cation transporter hOCT1 and their functional consequences. Pharmacogenetics 12:591–595

Kerner J, Hoppel C (1998) Genetic disorders of carnitine metabolism and their nutritional management. Annu Rev Nutr 18:179–206

Khamdang S, Takeda M, Noshiro R, Narikawa S, Enomoto A, Anzai N, Piyachaturawat P, Endou H (2002) Interactions of human organic anion transporters and human organic cation transporters with nonsteroidal anti-inflammatory drugs. J Pharmacol Exp Ther 303:534–539

Kilty JE, Lorang D, Amara SG (1991) Cloning and expression of a cocaine-sensitive rat dopamine transporter. Science 254:578–579

Kimelblatt BJ, Cerra F B, Calleri G, Berg MJ, McMillen MA, Schentag JJ (1980) Dose and serum concentration relationships in cimetidine-associated mental confusion. Gastroenterology 78:791–795

Kimura H, Takeda M, Narikawa S, Enomoto A, Ichida K, Endou H (2002) Human organic anion transporters and human organic cation transporters mediate renal transport of prostaglandins. J Pharmacol Exp Ther 301:293–298

Kinsella JL, Holohan PD, Pessah NI, Ross CR (1979) Transport of organic ions in renal cortical luminal and antiluminal membrane vesicles. J Pharmacol Exp Ther 209:443–450

Kippenberger AG, Palmer DJ, Comer AM, Lipski J, Burton LD, Christie DL (1999) Localization of the noradrenaline transporter in rat adrenal medulla and PC12 cells: evidence for its association with secretory granules in PC12 cells. J Neurochem 73:1024–1032

Kirkpatrick CJ, Bittinger F, Unger RE, Kriegsmann J, Kilbinger H, Wessler I (2001) The nonneuronal cholinergic system in the endothelium: evidence and possible pathobiological significance. Jpn J Pharmacol 85:24–28

Kobayashi Y, Okuda T, Fujioka Y, Matsumura G, Nishimura Y, Haga T (2002) Distribution of the high-affinity choline transporter in the human and macaque monkey spinal cord. Neurosci Lett 317:25–28

Koehler MR, Wissinger B, Gorboulev V, Koepsell H, Schmid M (1997) The two human organic cation transporter genes SLC22A1 and SLC22A2 are located on chromosome 6q26. Cytogenet Cell Genet 79:198–200

Koepsell H (1998) Organic cation transporters in intestine, kidney, liver, and brain. Annu Rev Physiol 60:243–266

Koepsell H, Gorboulev V, Arndt P (1999) Molecular pharmacology of organic cation transporters in kidney. J Membrane Biol 167:103–117

Koizumi A, Nozaki J-I, Ohura T, Kayo T, Wada Y, Nezu J-I, Ohashi R, Tamai I, Shoji Y, Takada G, Kibira S, Matsuishi T, Tsuji A (1999) Genetic epidemiology of the carnitine transporter OCTN2 gene in a Japanese population and phenotypic characterization in Japanese pedigrees with primary systemic carnitine deficiency. Hum Mol Genet 8:2247–2254

Koizumi T, Nikaido H, Hayakawa J, Nonomura A, Yoneda T (1988) Infantile disease with microvesicular fatty infiltration of viscera spontaneously occurring in the C3H-H-2° strain of mouse with similarities to Reye's syndrome. Lab Anim 22:83–87

Kristufek D, Rudorfer W, Pifl C, Huck S (2002) Organic cation transporter mRNA and function in the rat superior cervical ganglion. J Physiol 543:117–134

Kuhar MJ, Murrin LC (1978) Sodium-dependent, high affinity choline uptake. J Neurochem 30:15–21

Kuhl DE, Koeppe RA, Minoshima S, Snyder SE, Ficaro EP, Foster NL, Frey KA, Kilbourn MR (1999) In vivo mapping of cerebral acetylcholinesterase activity in aging and Alzheimer's disease. Neurology 52:691–699

Kullak-Ublick GA, Hagenbuch B, Stieger B, Wolkoff AW, Meier P J (1994) Functional characterization of the basolateral rat liver organic anion transporting polypeptide. Hepatology 20:411–416

Kullak-Ublick GA, Hagenbuch B, Stieger B, Schteingart CD, Hofmann AF, Wolkoff AW, Meier PJ (1995) Molecular and functional characterization of an organic anion transporting polypeptide cloned from human liver. Gastroenterology 109:1274–1282

Kullak-Ublick GA, Beuers U, Paumgartner G (1996) Molecular and functional characterization of bile acid transport in human hepatoblastoma HepG2 cells. Hepatology 23:1053–1060

Kullak-Ublick GA, Glasa J, Böker C, Oswald M, Grützner U, Hagenbuch B, Stieger B, Meier PJ, Beuers U, Kramer W, Wess G, Paumgartner G (1997) Chlorambucil-taurocholate is transported by bile acid carriers expressed in human hepatocellular carcinomas. Gastroenterology 113:1295–1305

Kullak-Ublick GA, Ismair M G, Stieger B, Landmann L, Huber R, Pizzagalli F, Fattinger K, Meier PJ, Hagenbuch B (2001) Organic anion-transporting polypeptide B (OATP-B) and its functional comparison with three other OATPs of human liver. Gastroenterology 120:525–533

Kusuhara H, Sekine T, Utsunomiya-Tate N, Tsuda M, Kojima R, Cha SH, Sugiyama Y, Kanai Y, Endou H (1999) Molecular cloning and characterization of a new multispecific organic anion transporter from rat brain. J Biol Chem 274:13675–13680

Kuwajima M, Kono N, Horiuchi M, Imamura Y, Ono A, Inui Y, Kawata S, Koizumi T, Hayakawa J, Saheki T (1991) Animal model of systemic carnitine deficiency: Analysis in C3H-H-2 degrees strain of mouse associated with juvenile visceral steatosis. Biochem Biophys Res Commun 174:1090–1094

Kwong SC, Brubacher J (1998) Phenformin and lactic acidosis: a case report and review. J Emerg Med 16:881–886

Lamhonwah A-M, Olpin SE, Pollitt RJ, Vianey-Saban C, Divry P, Guffon N, Besley GTN, Onizuka R, De Meirleir LJ, Cvitanovic-Sojat L, Baric I, Dionisi-Vici C, Fumic K, Maradin M, Tein I (2002) Novel OCTN2 mutations: no genotype-phenotype correlations: early carnitine therapy prevents cardiomyopathy. Am J Med Genet 111:271–284

Leabman MK, Huang CC, Kawamoto M, Johns SJ, Stryke D, Ferrin TE, DeYoung J, Taylor T, Clark AG, Herskowitz I, Giacomini KM (2002) Polymorphisms in a human kidney xenobiotic transporter, OCT2, exhibit altered function. Pharmacogenetics 12:395–405

Lecureur V, Guillouzo A, Fardel O (1998) Differential expression of the polyspecific drug transporter OCT1 in rat hepatocarcinoma cells. Cancer Lett 126:227–233

Lesch KP, Wolozin BL, Estler HC, Murphy DL, Riederer P (1993) Isolation of a cDNA encoding the human brain serotonin transporter. J Neural Transm Gen Sect 91:67–72

Lips KS, Pfeil U, Haberberger RV, Kummer W (2002) Localization of the high-affinity choline transporter-1 in the rat skeletal motor unit. Cell Tissue Res 307:275–280

Lockman PR, Allen DD (2002) The transport of choline. Drug Dev Ind Pharm 28:749–771

Lu K-M, Nishimori H, Nakamura Y, Shima K, Kuwajima M (1998) A missense mutation of mouse OCTN2, a sodium-dependent carnitine cotransporter, in the juvenile visceral steatosis mouse. Biochem Biophys Res Commun 252:590–594

Marger MD, Saier MH, Jr. (1993) A major superfamily of transmembrane facilitators that catalyse uniport, symport, and antiport. Trends Biochem Sci 18:13–20

Martel F, Keating E, Calhau C, Gründemann D, Schömig E, Azevedo I (2001) Regulation of human extraneuronal monoamine transporter (hEMT) expressed in HEK293 cells by intracellular second messenger systems. Naunyn Schmiedebergs Arch Pharmacol 364:487–495

Masson J, Sagne C, Hamon M, El Mestikawy S (1999) Neurotransmitter transporters in the central nervous system. Pharmacol Rev 51:439–464

Matsuoka M, Igisu H (1993) Comparison of the effects of L-carnitine, D-carnitine and acetyl-L- carnitine on the neurotoxicity of ammonia. Biochem Pharmacol 46:159–164

Mayatepek E, Nezu J, Tamai I, Oku A, Katsura M, Shimane M, Tsuji A (1999) Two novel missense mutations of the OCTN2 gene (W283R and V446F) in a patient with primary systemic carnitine deficiency. Hum Mutat 15:118

McCleskey EW, Almers W (1985) The Ca channel in skeletal muscle is a large pore. Proc Natl Acad Sci USA 82:7149–7153

McKinney TD, Kunnemann ME (1985) Procainamide transport in rabbit renal cortical brush border membrane vesicles. Am J Physiol Renal Physiol 249:F532–F541

Mehrens T, Lelleck S, Çetinkaya I, Knollmann M, Hohage H, Gorboulev V, Boknik P, Koepsell H, Schlatter E (2000) The affinity of the organic cation transporter rOCT1 is increased by protein kinase C-dependent phosphorylation. J Am Soc Nephrol 11:1216–1224

Meijer DKF, Smit JW, Müller M (1997) Hepatobiliary elimination of cationic drugs: the role of P-glycoproteins and other ATP-dependent transporters. Adv Drug Deliv Rev 25:159–200

Meyer-Wentrup F, Karbach U, Gorboulev V, Arndt P, Koepsell H (1998) Membrane localization of the electrogenic cation transporter rOCT1 in rat liver. Biochem Biophys Res Commun 248:673–678

Mezey E, Eisenhofer G, Harta G, Hansson S, Gould L, Hunyady B, Hoffman BJ (1996) A novel nonneuronal catecholaminergic system: exocrine pancreas synthesizes and releases dopamine. Proc Natl Acad Sci USA 93:10377–10382

Mezey E, Eisenhofer G, Hansson S, Hunyady B, Hoffman BJ (1998) Dopamine produced by the stomach may act as a paracrine/autocrine hormone in the rat. Neuroendocrinology 67:336–348

Mezey E, Eisenhofer G, Hansson S, Harta G, Hoffman BJ, Gallatz K, Palkovits M, Hunyady B (1999) Nonneuronal dopamine in the gastrointestinal system. Clin Exp Pharmacol Physiol 26 :S14–S22

Mooslehner KA, Allen ND (1999) Cloning of the mouse organic cation transporter 2 gene, Slc22a2, from an enhancer-trap transgene integration locus. Mamm Genome 10:218–224

Moseley RH, Morrissette J, Johnson TR (1990) Transport of N^1-methylnicotinamide by organic cation-proton exchange in rat liver membrane vesicles. Am J Physiol Gastroenterol 259:G973–G982

Moseley RH, Jarose SM, Permoad P (1992a) Organic cation transport by rat liver plasma membrane vesicles: studies with TEA. Am J Physiol Gastroenterol Liver Physiol 263:G775–G785

Moseley RH, Vashi P G, Jarose SM, Dickinson CJ, Permoad PA (1992b) Thiamine transport by basolateral rat liver plasma membrane vesicles. Gastroenterology 103:1056–1065

Motohashi H, Sakurai Y, Saito H, Masuda S, Urakami Y, Goto M, Fukatsu A, Ogawa O, Inui KI (2002) Gene expression levels and immunolocalization of organic ion transporters in the human kidney. J Am Soc Nephrol 13:866–874

Nagao M, Misawa H, Kato S, Hirai S (1998) Loss of cholinergic synapses on the spinal motor neurons of amyotrophic lateral sclerosis. J Neuropathol Exp Neurol 57:329–333

Nagel G, Volk C, Friedrich T, Ulzheimer J C, Bamberg E, Koepsell H (1997) A reevaluation of substrate specificity of the rat cation transporter rOCT1. J Biol Chem 272:31953–31956

Nestler JE (2001) Metformin and the polycystic ovary syndrome. J Clin Endocrinol Metab 86:1430

Nezu J, Tamai I, Oku A, Ohashi R, Yabuuchi H, Hashimoto N, Nikaido H, Sai Y, Koizumi A, Shoji Y, Takada G, Matsuishi T, Yoshino M, Kato H, Ohura T, Tsujimoto G, Hayakawa J, Shimane M, Tsuji A (1999) Primary systemic carnitine deficiency is caused by mutations in a gene encoding sodium ion-dependent carnitine transporter. Nat Genet 21:91–94

Nguyen TT, Tseng Y T, McGonnigal B, Stabila JP, Worrell L A, Saha S, Padbury JF (1999) Placental biogenic amine transporters: in vivo function, regulation and pathobiological significance. Placenta 20:3–11

Nicholas TE, Strum J M, Angelo LS, Junod AF (1974) Site and mechanism of uptake of ^{3}H-norepinephrine by isolated perfused rat lungs. Circ Res 35:670–680

Nishiwaki T, Daigo Y, Tamari M, Fujii Y, Nakamura Y (1998) Molecular cloning, mapping, and characterization of two novel human genes, ORCTL3 and ORCTL4, bearing homology to organic-cation transporters. Cytogenet Cell Genet 83:251–255

O'Regan S, Traiffort E, Ruat M, Cha N, Compaore D, Meunier FM (2000) An electric lobe suppressor for a yeast choline transport mutation belongs to a new family of transporter-like proteins. Proc Natl Acad Sci USA 97:1835–1840

Ohashi R, Tamai I, Yabuuchi H, Nezu J-I, Oku A, Sai Y, Shimane M and Tsuji A (1999) Na(+)-dependent carnitine transport by organic cation transporter (OCTN2): its pharmacological and toxicological relevance. J Pharmacol Exp Ther 291:778–784

Ohashi R, Tamai I, Nezu J-I, Nikaido H, Hashimoto N, Oku A, Sai Y, Shimane M, Tsuji A (2001) Molecular and physiological evidence for multifunctionality of carnitine/organic cation transporter OCTN2. Mol Pharmacol 59:358–366

Ohashi R, Tamai I, Inano A, Katsura M, Sai Y, Nezu J, Tsuji A (2002) Studies on functional sites of organic cation/carnitine transporter OCTN2 (SLC22A5) using a Ser467Cys mutant protein. J Pharmacol Exp Ther 302:1286–1294

Oishi K, Hirai T, Gelb BD, Diaz GA (2001) Slc19a2: Cloning and characterization of the murine thiamin transporter cDNA and genomic sequence, the orthologue of the human TRMA gene. Mol Genet Metab 73:149–159

Okuda M, Saito H, Urakami Y, Takano M, Inui K-I (1996) cDNA cloning and functional expression of a novel rat kidney organic cation transporter, OCT2. Biochem Biophys Res Commun 224:500–507

Okuda M, Urakami Y, Saito H, Inui K-I (1999) Molecular mechanisms of organic cation transport in OCT2-expressing *Xenopus* oocytes. Biochim Biophys Acta 1417:224–231

Okuda T, Haga T (2000) Functional characterization of the human high-affinity choline transporter. FEBS Lett 484:92–97

Okuda T, Haga T, Kanai Y, Endou H, Ishihara T, Katsura I (2000) Identification and characterization of the high-affinity choline transporter. Nat Neurosci 3:120–125

Okuda T, Okamura M, Kaitsuka C, Haga T, Gurwitz D (2002) Single nucleotide polymorphism of the human high-affinity choline transporter alters transport rate. J Biol Chem 277:45314–45322

Ott RJ, Hui AC, Yuan G, Giacomini KM (1991) Organic cation transport in human renal brush-border membrane vesicles. Am J Physiol Renal Physiol 261:F443–F451

Owen MR, Doran E, Halestrap AP (2000) Evidence that metformin exerts its antidiabetic effects through inhibition of complex 1 of the mitochondrial respiratory chain. Biochem J 348:607–614

Pacholczyk T, Blakely RD, Amara SG (1991) Expression cloning of a cocaine- and antidepressant-sensitive human noradrenaline transporter. Nature 350:350–354

Pan BF, Sweet DH, Pritchard JB, Chen R, Nelson JA (1999) A transfected cell model for the renal toxin transporter, rOCT2. Toxicol Sci 47:181–186

Pao SS, Paulsen IT, Saier MH, Jr. (1998) Major facilitator superfamily. Microbiol Mol Biol Rev 62:1–34

Parsons SM, Prior C, Marshall IG (1993) Acetylcholine transport, storage, and release. Int Rev Neurobiol 35:279–390

Pettegrew JW, Klunk WE, Panchalingam K, Kanfer JN, McClure R (1995) Clinical and neurochemical effects of acetyl-L-carnitine in Alzheimer's disease. Neurobiol Aging16:1–4

Phillips JK, Dubey R, Sesiashvilvi E, Takeda M, Christie DL and Lipski J (2001) Differential expression of the noradrenaline transporter in adrenergic chromaffin cells, ganglion cells and nerve fibres of the rat adrenal medulla. J Chem Neuroanat 21:95–104

Pietig G, Mehrens T, Hirsch JR, Çetinkaya I, Piechota H, Schlatter E (2001) Properties and regulation of organic cation transport in freshly isolated human proximal tubules. J Biol Chem 276:33741–33746

Prasad PD, Huang W, Ramamoorthy S, Carter AL, Leibach FH, Ganapathy V (1996) Sodium-dependent carnitine transport in human placental choriocarcinoma cells. Biochim Biophys Acta 1284:109–117

Rajgopal A, Edmondnson A, Goldman ID, Zhao R (2001) *SLC19A3* encodes a second thiamine transporter ThTr2. Biochim Biophys Acta 1537:175–178

Ramamoorthy S, Bauman AL, Moore KR, Han H, Yang-Feng T, Chang AS, Ganapathy V, Blakely RD (1993) Antidepressant- and cocaine-sensitive human serotonin transporter: molecular cloning, expression, and chromosomal localization. Proc Natl Acad Sci USA 90:2542–2546

Rebouche CJ and Mack DL (1984) Sodium gradient-stimulated transport of L-carnitine into renal brush border membrane vesicles: kinetics, specificity, and regulation by dietary carnitine. Arch Biochem Biophys 235:393–402

Reichel C, Gao B, Van Montfoort J, Cattori V, Rahner C, Hagenbuch B, Stieger B, Kamisako T, Meier PJ (1999) Localization and function of the organic anion-transporting polypeptide Oatp2 in rat liver. Gastroenterology 117:688–695

Reid G, Wolff NA, Dautzenberg FM, Burckhard G (1998) Cloning of a human renal *p*-aminohippurate transporter, hROAT1. Kidney Blood Press Res 21:233–237

Rindi G, Laforenza U (2000) Thiamine intestinal transport and related issues: recent aspects. Proc Soc Exp Biol Med 224:246–255

Roch-Ramel F, Besseghir K, Murer H (1992) Renal excretion and tubular transport of organic anions and cations. In Windhager EE (ed) Handbook of physiology (a critical, comprehensive presentation of physiological knowledge and concepts). Oxford University Press, New York, Oxford, pp 2189–2262

Roque AS, Prasad P D, Bhatia JS, Leibach FH, Ganapathy V (1996) Sodium-dependent high-affinity binding of carnitine to human placental brush border membranes. Biochim Biophys Acta 1282:274–282

Russ H, Staust K, Martel F, Gliese M, Schömig E (1996) The extraneuronal transporter for monoamine transmitters exists in cells derived from human central nervous system glia. Eur J Neurosc 8:1256–1264

Sakaeda T, Nakamura T, Okumura K (2002) MDR1 genotype-related pharmacokinetics and pharmacodynamics. Biol Pharm Bull 25:1391–1400

Schentag JJ, Cerra FB, Calleri G, DeGlopper E, Rose JQ, Bernhard H (1979) Pharmacokinetic and clinical studies in patients with cimetidine-associated mental confusion. Lancet 1:177–181

Schlatter E, Monnich V, Çetinkaya I, Mehrens T, Ciarimboli G, Hirsch JR, Popp C, Koepsell H (2002) The organic cation transporters rOCT1 and hOCT2 are inhibited by cGMP. J Membr Biol 189:237–244

Schmitt A, Mössner R, Gossmann A, Fischer I G, Gorboulev V, Murphy DL, Koepsell H, Lesch KP (2003) An organic cation transporter capable of transporting serotonin is upregulated in serotonin transporter deficient-mice. J Neurosci Res (in press)

Schömig E, Schönfeld C-L (1990) Extraneuronal noradrenaline transport (uptake$_2$) in a human cell line (Caki-1 cells). Naunyn Schmiedeberg Arch Pharmacol 341:404–410

Schömig E, Spitzenberger F, Engelhardt M, Martel F, Örding N, Gründemann D (1998) Molecular cloning and characterization of two novel transport proteins from rat kidney. FEBS Letters 425:79–86

Schweifer N, Barlow DP (1996) The *Lx1* gene maps to mouse chromosome 17 and codes for a protein that is homologous to glucose and polyspecific transmembrane transporters. Mamm Genome 7:735–740

Sekine T, Watanabe N, Hosoyamada M, Kanai Y, Endou H (1997) Expression cloning and characterization of a novel multispecific organic anion transporter. J Biol Chem 272:18526–18529

Sekine T, Kusuhara H, Utsunomiya-Tate N, Tsuda M, Sugiyama Y, Kanai Y, Endou H (1998) Molecular cloning and characterization of high-affinity carnitine transporter from rat intestine. Biochem Biophys Res Commun 251:586–591

Sekine T, Cha S H, Endou H (2000) The multispecific organic anion transporter (OAT) family. Pflügers Arch 440:337–350

Seth P, Wu X, Huang W, Leibach FH, Ganapathy V (1999) Mutations in novel organic cation transporter (OCTN2), an organic cation/carnitine transporter, with differential effects on the organic cation transport function and the carnitine transport function. J Biol Chem 274:33388–33392

Shi X, Bai S, Ford AC, Burk RD, Jacquemin E, Hagenbuch B, Meier PJ, Wolkoff AW (1995) Stable inducible expression of a functional rat liver organic anion transport protein in HeLa cells. J Biol Chem 270:25591–25595

Shimada S, Kitayama S, Lin C-L, Patel A, Nanthakumar E, Gregor P, Kuhar M, Uhl G (1991) Cloning and expression of a cocaine-sensitive dopamine transporter complementary DNA. Science 254:576–578

Shu Y, Bello CL, Mangravite LM, Feng B, Giacomini KM (2001) Functional characteristics and steroid hormone-mediated regulation of an organic cation transporter in Madin-Darby canine kidney cells. J Pharmacol Exp Ther 299:392–398

Simon JR, Kuhar MG (1975) Impulse-flow regulation of high affinity choline uptake in brain cholinergic nerve terminals. Nature 255:162–163

Slitt AL, Cherrington NJ, Hartley DP, Leazer TM, Klaassen CD (2002) Tissue distribution and renal developmental changes in rat organic cation transporter mRNA levels. Drug Metab Dispos 30:212–219

Smit JW, Schinkel AH, Weert B, Meijer DKF (1998) Hepatobiliary and intestinal clearance of amphiphilic cationic drugs in mice in which both *mdr1a* and *mdr1b* genes have been disrupted. Br J Pharmacol 124:416–424

Sokol PP, Holohan PD, Ross CR (1985) Electroneutral transport of organic cations in canine renal brush border membrane vesicles (BBMV). J Pharmacol Exp Ther 233:694–699

Spangeus A, El-Salhy M (2001) Myenteric plexus of obese diabetic mice (an animal model of human type 2 diabetes). Histol Histopathol 16:159–165

Stanley CA, DeLeeuw S, Coates PM, Vianey-Liaud C, Divry P, Bonnefont JP, Saudubray JM, Haymond M, Trefz FK, Breningstall GN (1991) Chronic cardiomyopathy and weakness or acute coma in children with a defect in carnitine uptake. Ann Neurol 30:709–716

Steffen V, Santiago M, de la Cruz CP, Revilla E, Machado A, Cano J (1995) Effect of intraventricular injection of MPP: protection by acetyl-L-carnitine. Hum Exp Toxicol 14:865–871

Stieger B, O'Neill B, Krähenbühl S (1995) Characterization of L-carnitine transport by rat kidney brush-border-membrane vesicles. Biochem J 309:643–647

Streich S, Brüss M, Bönisch H (1996) Expression of the extraneuronal monoamine transporter (uptake$_2$) in human glioma cells. Naunyn-Schmiedeberg's Arch Pharmacol 353:328–333

Sugawara-Yokoo M, Urakami Y, Koyama H, Fujikura K, Masuda S, Saito H, Naruse T, Inui K-I, Takata K (2000) Differential localization of organic cation transporters rOCT1 and rOCT2 in the basolateral membrane of rat kidney proximal tubules. Histochem Cell Biol 114:175–180

Sun W, Wu RR, van Poelje PD, Erion MD (2001) Isolation of a family of organic anion transporters from human liver and kidney. Biochem Biophys Res Commun 283:417–422.

Suzuki M, Desmond TJ, Albin RL, Frey KA (2002) Cholinergic vesicular transporters in progressive supranuclear palsy. Neurology 58:1013–1018

Sweet DH, Pritchard JB (1999a) rOCT2 is a basolateral potential-driven carrier, not an organic cation/proton exchanger. Am J Physiol Renal Physiol 277:F890–F898

Sweet DH, Pritchard JB (1999b) The molecular biology of renal organic anion and organic cation transporters. Cell Biochem Biophys 31:89–118

Sweet DH, Wolff NA, Pritchard JB (1997) Expression cloning and characterization of ROAT1, the basolateral organic anion transporter in rat kidney. J Biol Chem 272:30088–30095

Sweet DH, Miller DS, Pritchard JB (2001) Ventricular choline transport: a role for organic cation transporter 2 expressed in choroid plexus. J Biol Chem 276:41611–41619

Sweet DH, Chan LMS, Walden R, Yang X-P, Miller DS, Pritchard JB (2003) Organic anion transporter 3 (*Slc22a8*) is a dicarboxylate exchanger indirectly coupled to the Na^+ gradient. Am J Physiol Renal Physiol (in press)

Takano M, Inui K-I, Okano T, Saito H, Hori R (1984) Carrier-mediated transport systems of TEA in rat renal brush-border and basolateral membrane vesicles. Biochim Biophys Acta 773:113–124

Takano M, Katsura T, Tomita Y, Yasuhara M, Hori R (1993) Transport mechanism of choline in rat renal brush-border membrane. Biol Pharm Bull 16:889–894

Takeda M, Khamdang S, Narikawa S, Kimura H, Kobayashi Y, Yamamoto T, Cha SH, Sekine T, Endou H (2002) Human organic anion transporters and human organic cation transporters mediate renal antiviral transport. J Pharmacol Exp Ther 300:918–924

Tamai I, Yabuuchi H, Nezu J-I, Sai Y, Oku A, Shimane M, Tsuji A (1997) Cloning and characterization of a novel human pH-dependent organic cation transporter, OCTN1. FEBS Letters 419:107–111

Tamai I, Ohashi R, Nezu J-I, Yabuuchi H, Oku A, Shimane M, Sai Y, Tsuji A (1998) Molecular and functional identification of sodium ion-dependent, high affinity human carnitine transporter OCTN2. J Biol Chem 273:20378–20382

Tamai I, Ohashi R, Nezu J-I, Sai Y, Kobayashi D, Oku A, Shimane M, Tsuji A (2000) Molecular and functional characterization of organic cation/carnitine transporter family in mice. J Biol Chem 275:40064–40072

Tamai I, China K, Sai Y, Kobayashi D, Nezu J-I, Kawahara E, Tsuji A (2001) Na(+)-coupled transport of L-carnitine via high-affinity carnitine transporter OCTN2 and its subcellular localization in kidney. Biochim Biophys Acta 1512:273–284

Tang NLS, Ganapathy V, Wu X, Hui J, Seth P, Yuen PMP, Fok TF, Hjelm NM (1999) Mutations of OCTN2, an organic cation/carnitine transporter, lead to deficient cellular carnitine uptake in primary carnitine deficiency. Hum Mol Genet 8:655–660

Tanigawara Y, Okamura N, Hirai M, Yasuhara M, Ueda K, Kioka N, Komano T, Hori R (1992) Transport of digoxin by human P-glycoprotein expressed in a porcine kidney epithelial cell line (LLC-PK_1). J Pharmacol Exp Ther 263:840–845

Tein I, De Vivo DC, Bierman F, Pulver P, De Meirleir LJ, Cvitanovic-Sojat L, Pagon R A, Bertini E, Dionisi-Vici C, Servidei S. (1990) Impaired skin fibroblast carnitine uptake in primary systemic carnitine deficiency manifested by childhood carnitine-responsive cardiomyopathy. Pediatr Res 28:247–255

Terashita S, Dresser MJ, Zhang L, Gray AT, Yost SC, Giacomini KM (1998) Molecular cloning and functional expression of a rabbit renal organic cation transporter. Biochim Biophys Acta 1369:1–6

Thiebaut F, Tsuruo T, Hamada H, Gottesman MM, Pastan I, Willingham MC (1987) Cellular localization of the multidrug-resistance gene product P-glycoprotein in normal human tissues. Proc Natl Acad Sci USA 84:7735–7738

Thiebaut F, Tsuruo T, Hamada H, Gottesman MM, Pastan I, Willingham MC (1989) Immunohistochemical localization in normal tissues of different epitopes in the multidrug transport protein P170: evidence for localization in brain capillaries and crossreactivity of one antibody with a muscle protein. J Histochem Cytochem 37:159–164

Tomomura M, Imamura Y, Horiuchi M, Koizumi T, Nikaido H, Hayakawa J, Saheki T (1992) Abnormal expression of urea cycle enzyme genes in juvenile visceral steatosis (*jvs*) mice. Biochim Biophys Acta 1138:167–171

Treem WR, Stanley CA, Finegold DN, Hale DE, Coates PM (1988) Primary carnitine deficiency due to a failure of carnitine transport in kidney, muscle, and fibroblasts. N Engl J Med 319:1331–1336

Trendelenburg U (1988) The extraneural uptake and metabolism of catecholamines. In Trendelenburg U, Weiner N (eds) Handbook of experimental pharmacology 90. Catecholamines I. Springer-Verlag, Berlin, pp 279–319

Turnheim K, Lauterbach FO (1977) Absorption and secretion of monoquaternary ammonium compounds by the isolated intestinal mucosa. Biochem Pharmac 26:99–108

Ueda K, Okamura N, Hirai M, Tanigawara Y, Saeki T, Kioka N, Komano T, Hori R (1992) Human P-glycoprotein transports cortisol, aldosterone, and dexamethasone, but not progesterone. J Biol Chem 267:24248–24252

Ullrich KJ (1994) Specificity of transporters for "organic anions" and "organic cations" in the kidney. Biochim Biophys Acta 1197:45–62

Ullrich KJ, Rumrich G (1996) Luminal transport system for $choline^+$ in relation to the other organic cation transport systems in the rat proximal tubule. Kinetics, specificity: alkyl/arylamines, alkylamines with OH, O, SH, NH_2, ROCO, RSCO, and H_2PO_4-groups, methylaminostyryl, rhodamine, acridine, phenanthrene, and cyanine compounds. Pflügers Arch 432:471–485

Urakami Y, Okuda M, Masuda S, Saito H, Inui K-I (1998) Functional characteristics and membrane local-
    ization of rat multispecific organic cation transporters, OCT1 and OCT2, mediating tubular secretion
    of cationic drugs. J Pharmacol Exp Ther 287:800–805
Urakami Y, Nakamura N, Takahashi K, Okuda M, Saito H, Hashimoto Y, Inui K-I (1999) Gender differ-
    ences in expression of organic cation transporter OCT2 in rat kidney. FEBS Lett 461:339–342
Urakami Y, Okuda M, Saito H, Inui K-I (2000) Hormonal regulation of organic cation rransporter OCT2
    expression in rat kidney. FEBS Lett 473:173–176
Urakami Y, Okuda M, Masuda S, Akazawa M, Saito H, Inui K-I (2001) Distinct characteristics of organic
    cation transporters, OCT1 and OCT2, in the basolateral membrane of renal tubules. Pharm Res
    18:1528–1534
Urakami Y, Akazawa M, Saito H, Okuda M, Inui K-I (2002) cDNA cloning, functional characterization,
    and tissue distribution of an alternatively spliced variant of organic cation transporter hOCT2 predomi-
    nantly expressed in the human kidney. J Am Soc Nephrol 13:1703–1710
Van Montfoort J E, Hagenbuch B, Fattinger KE, Müller M, Groothuis GMM, Meijer DKF, Meier PJ (1999)
    Polyspecific organic anion transporting polypeptides mediate hepatic uptake of amphipathic type II or-
    ganic cations. J Pharmacol Exp Ther 291:147–152
Van Montfoort JE, Müller M, Groothuis GMM, Meijer DKF, Koepsell H, Meier PJ (2001) Comparison of
    "type I" and "type II" organic cation transport by organic cation transporters and organic anion-trans-
    porting polypeptides. J Pharmacol Exp Ther 298:110–115
Varoqui H, Erickson JD (1996) Active transport of acetylcholine by the human vesicular acetylcholine
    transporter. J Biol Chem 271:27229–27232
Vaz FM, Scholte HR, Ruiter J, Hussaarts-Odijk LM, Pereira RR, Schweitzer S, de Klerk JBC, Waterham
    HR, Wanders RJA (1999) Identification of two novel mutations in OCTN2 of three patients with sys-
    temic carnitine deficiency. Hum Genet 105:157–161
Velazquez EM, Mendoza S, Hamer T, Sosa F, Glueck CJ (1994) Metformin therapy in polycystic ovary
    syndrome reduces hyperinsulinemia, insulin resistance, hyperandrogenemia, and systolic blood pres-
    sure, while facilitating normal menses and pregnancy. Metabolism 43:647–654
Verhaagh S, Schweifer N, Barlow DP, Zwart R (1999) Cloning of the mouse and human solute carrier 22a3
    (Slc22a3/SLC22A3) identifies a conserved cluster of three organic cation transporters on mouse chro-
    mosome 17 and human 6q26-q27. Genomics 55:209–218
Vialli M (1966) Histology of the enterochromaffin cell system. In: Erspamer V (ed) Handbook of experi-
    mental pharmacology: 5-hydroxytryptamine and related indolealkyklaminase 19. Springer, New York,
    pp 1–65
Wagner CA, Lükewille U, Kaltenbach S, Moschen I, Bröer A, Risler T, Bröer S, Lang F (2000) Functional
    and pharmacological characterization of the human Na^+/carnitine cotransporter hOCTN2. Am J Phy-
    siol Renal Physiol 279:F584–F591
Walker JK, Gainetdinov RR, Mangel AW, Caron MG and Shetzline MA (2000) Mice lacking the dopamine
    transporter display altered regulation of distal colonic motility. Am J Physiol Gastrointest Liver Phy-
    siol 279:G311–G318
Walker PS, Donovan JA, Van Ness BG, Fellows RE, Pessin JE (1988) Glucose-dependent regulation of
    glucose transport activity, protein, and mRNA in primary cultures of rat brain glial cells. J Biol Chem
    263:15594–15601
Wang DS, Jonker JW, Kato Y, Kusuhara H, Schinkel AH, Sugiyama Y (2002) Involvement of organic cation
    transporter 1 in hepatic and intestinal distribution of metformin. J Pharmacol Exp Ther 302:510–515
Wang Y, Ye J, Ganapathy V, Longo N (1999) Mutations in the organic cation/carnitine transporter OCTN2
    in primary carnitine deficiency. Proc Natl Acad Sci USA 96:2356–2360
Wang Y, Taroni F, Garavaglia B, Longo N (2000a) Functional analysis of mutations in the OCTN2 trans-
    porter causing primary carnitine deficiency: lack of genotype-phenotype correlation. Hum Mutat
    16:401–407
Wang Y, Kelly MA, Cowan TM, Longo N (2000b) A missense mutation in the OCTN2 gene associated
    with residual carnitine transport activity. Hum Mutat 15:238–245
Wang Y, Meadows TA, Longo N (2000c) Abnormal sodium stimulation of carnitine transport in primary
    carnitine deficiency. J Biol Chem 275:20782–20786
Wang Y, Korman SH, Ye J, Gargus JJ, Gutman A, Taroni F, Garavaglia B, Longo N (2001) Phenotype and
    genotype variation in primary carnitine deficiency. Genet Med 3:387–392
Wessler I, Kirkpatrick CJ, Racke K (1999) The cholinergic "pitfall": acetylcholine, a universal cell mole-
    cule in biological systems, including humans. Clin Exp Pharmacol Physiol 26:198–205
Wessler I, Roth E, Deutsch C, Brockerhoff P, Bittinger F, Kirkpatrick CJ, Kilbinger H (2001a) Release of
    nonneuronal acetylcholine from the isolated human placenta is mediated by organic cation trans-
    porters. Br J Pharmacol 134:951–956

Wessler I, Roth E, Schwarze S, Weikel W, Bittinger F, Kirkpatrick CJ, Kilbinger H (2001b) Release of nonneuronal acetylcholine from the human placenta: difference to neuronal acetylcholine. Naunyn Schmiedebergs Arch Pharmacol 364:205–212

Wolff NA, Werner A, Burkhardt S, Burckhardt G (1997) Expression cloning and characterization of a renal organic anion transporter from winter flounder. FEBS Letters 417:287–291

Wolff NA, Grünwald B, Friedrich B, Lang F, Godehardt S, Burckhardt G (2001) Cationic amino acids involved in dicarboxylate binding of the flounder renal organic anion transporter. J Am Soc Nephrol 12:2012–2018

Wright EM, Loo DDF, Panayotova-Heiermann M, Hirayama BA, Turk E, Eskandari S, Lam JT (1998) Structure and function of the Na$^+$/glucose cotransporter. Acta Physiol Scand 163:257–264

Wright SH (1985) Transport of N^1-methylnicotinamide across brush border membrane vesicles from rabbit kidney. Am J Physiol Renal Physiol 249:F903–F911

Wright SH, Wunz TM, Wunz TP (1992) A choline transporter in renal brush-border membrane vesicles: energetics and structural specificity. J Membrane Biol 126:51–65

Wu X, Prasad PD, Leibach FH and Ganapathy V (1998a) cDNA sequence, transport function, and genomic organization of human OCTN2, a new member of the organic cation transporter family. Biochem Biophys Res Commun 246:589–595

Wu X, Kekuda R, Huang W, Fei Y-J, Leibach FH, Chen J, Conway SJ, Ganapathy V (1998b) Identity of the organic cation transporter OCT3 as the extraneuronal monoamine transporter (uptake$_2$) and evidence for the expression of the transporter in the brain. J Biol Chem 273:32776–32786

Wu X, Huang W, Prasad PD, Seth P, Rajan DP, Leibach FH, Chen J, Conway SJ, Ganapathy V (1999) Functional characteristics and tissue distribution pattern of organic cation transporter 2 (OCTN2), an organic cation/carnitine transporter. J Pharmacol Exp Ther 290:1482–1492

Wu X, George RL, Huang W, Wang H, Conway SJ, Leibach FH, Ganapathy V (2000a) Structural and functional characteristics and tissue distribution pattern of rat OCTN1, an organic cation transporter, cloned from placenta. Biochim Biophys Acta 1466:315–327

Wu X, Huang W, Ganapathy ME, Wang H, Kekuda R, Conway SJ, Leibach FH, Ganapathy V (2000b) Structure, function, and regional distribution of the organic cation transporter OCT3 in the kidney. Am J Physiol Renal Physiol 279:F449–F458

Yabuuchi H, Tamai I, Nezu J-I, Sakamoto K, Oku A, Shimane M, Sai Y, Tsuji A (1999) Novel membrane transporter OCTN1 mediates multispecific, bidirectional, and pH-dependent transport of organic cations. J Pharmacol Exp Ther 289:768–773

Yamamura HI, Snyder SH (1972) Choline: high-affinity uptake by rat brain synaptosomes. Science 178:626–628

Yokogawa K, Yonekawa M, Tamai I, Ohashi R, Tatsumi Y, Higashi Y, Nomura M, Hashimoto N, Nikaido H, Hayakawa J, Nezu J, Oku A, Shimane M, Miyamoto K, Tsuji A (1999) Loss of wild-type carrier-mediated L-carnitine transport activity in hepatocytes of juvenile visceral steatosis mice. Hepatology 30:997–1001

Yoshioka K (1984) Some properties of the thiamine uptake system in isolated rat hepatocytes. Biochim Biophys Acta 778:201–209

Zhang L, Dresser MJ, Gray AT, Yost SC, Terashita S, Giacomini K M (1997a) Cloning and functional expression of a human liver organic cation transporter. Mol Pharmacol 51:913–921

Zhang L, Dresser MJ, Chun JK, Babbitt PC, Giacomini KM (1997b) Cloning and functional characterization of a rat renal organic cation transporter isoform (rOCT1A). J Biol Chem 272:16548–16554

Zhang L, Brett CM, Giacomini KM (1998a) Role of organic cation transporters in drug absorption and elimination. Annu Rev Pharmacol Toxicol 38:431–460

Zhang L, Schaner ME, Giacomini KM (1998b) Functional characterization of an organic cation transporter (hOCT1) in a transiently transfected human cell line (HeLa). J Pharmacol Exp Ther 286:354–361

Zhang L, Gorset W, Dresser MJ, Giacomini KM (1999) The interaction of n-tetra-alkylammonium compounds with a human organic cation transporter, hOCT1. J Pharmacol Exp Ther 288:1192–1198

Zhang L, Gorset W, Washington CB, Blaschke TF, Kroetz DL, Giacomini KM (2000) Interactions of HIV protease inhibitors with a human organic cation transporter in a mammalian expression system. Drug Metab Dispos 28:329–334

Zhang X, Evans KK, Wright SH (2002) Molecular cloning of rabbit organic cation transporter rbOCT2 and functional comparisons with rbOCT1. Am J Physiol Renal Physiol 283:F124–F133

Zwart R, Verhaagh S, Buitelaar M, Popp-Snijders C, Barlow DP (2001) Impaired activity of the extraneuronal monoamine transporter system known as uptake-2 in Orct3/Slc22a3-deficient mice. Mol Cell Biol 21:4188–4196

Rev Physiol Biochem Pharmacol (2003) 150:91–139
DOI 10.1007/s10254-003-0019-8

W. G. Wier · K. G. Morgan

α_1-Adrenergic signaling mechanisms in contraction of resistance arteries

Published online: 17 July 2003
© Springer-Verlag 2003

Abstract Our goal in this review is to provide a comprehensive, integrated view of the numerous signaling pathways that are activated by α_1-adrenoceptors and control actin-myosin interactions (i.e., crossbridge cycling and force generation) in mammalian arterial smooth muscle. These signaling pathways may be categorized broadly as leading either to thick (myosin) filament regulation or to thin (actin) filament regulation. Thick filament regulation encompasses both "Ca^{2+} activation" and "Ca^{2+}-sensitization" as it involves both activation of myosin light chain kinase (MLCK) by Ca^{2+}-calmodulin and regulation of myosin light chain phosphatase (MLCP) activity. With respect to Ca^{2+} activation, adrenergically induced Ca^{2+} transients in individual smooth muscle cells of intact arteries are now being shown by high resolution imaging to be sarcoplasmic reticulum-dependent asynchronous propagating Ca^{2+} waves. These waves differ from the spatially uniform increases in $[Ca^{2+}]$ previously assumed. Similarly, imaging during adrenergic activation has revealed the dynamic translocation, to membranes and other subcellular sites, of protein kinases (e.g., Ca^{2+}-activated protein kinases, PKCs) that are involved in regulation of MLCP and thus in "Ca^{2+} sensitization" of contraction. Thin filament regulation includes the possible disinhibition of actin-myosin interactions by phosphorylation of CaD, possibly by mitogen-activated protein (MAP) kinases that are also translocated during adrenergic activation. An hypothesis for the mechanisms of adrenergic activation of small arteries is advanced. This involves asynchronous Ca^{2+} waves in individual SMC, synchronous Ca^{2+} oscillations (at high levels of adrenergic activation), Ca^{2+} sparks, "Ca^{2+}-sensitization" by PKC and Rho-associated kinase (ROK), and thin filament mechanisms.

The French version of this article is available in the form of electronic supplementary material and can be obtained by using the Springer Link server located at http://dx.doi.org/10.1007/s10254-003-0019-8

W. G. Wier (✉)
Department of Physiology, School of Medicine, University of Maryland,
655 West Baltimore Street, Baltimore, MD, 21201, USA
e-mail: gwier001@umaryland.edu

K. G. Morgan
Boston Biomedical Research Institute,
64 Grove Street, Watertown, MA, 02472, USA

Abbreviations *2-APB:* 2-Aminoethoxydiphenylborate · *ABS-1:* Actin binding sequence no. 1 · *BK:* Large conductance potassium channel · *CaD:* Caldesmon · *CaM:* Calmodulin · *CaMKinase II:* Calmodulin kinase II · *CaP:* Calponin · *CICR:* Ca^{2+}-induced Ca^{2+} release · *CPA:* Cyclopiazonic acid · *CPI-17:* Protein kinase C-potentiated 17 kDa inhibitor protein · *2,4-DCB:* 2,4-Dichlorobenzamil · *DAG:* Diacylglycerol · *DHP:* Dihydropyridine · *DOG:* 1,2-Dioctanoyl-*sn*-glycerol · *ERK:* Extracellular-regulated kinase · *FDS:* Frequent discharge sites · *FRAP:* Fluorescence recovery after photobleaching · *FRET:* Fluorescence resonance energy transfer · *GEF:* Guanine nucleotide exchange factor · *GS17C:* Fluorophore peptide antagonist of caldesmon · *HA-1077:* 1-(5-Isoquinolinesulfonyl) homopiperazine, Di-HCl Salt · *IICR:* $InsP_3$ induced Ca^{2+} release · *ILK:* Integrin-linked kinase · *InsP$_3$R:* 1,4,5-Trisphosphate receptor · *IVC:* Inferior vena cava · *jCaTs:* Junctional calcium transients · *LC20:* 20,000 Da light chain of smooth muscle myosin · *M20:* Small noncatalytic subunit of myosin phosphatase · *M130:* Large noncatalytic subunit of myosin phosphatase · *MAP kinase:* Mitogen-activated protein kinase · *MEK:* MAPK kinase · *ML-9:* 1-(5-Chloronaphthalene-1-sulfonyl)-1H-hexahydro-1,4-diazepine hydrochloride · *MLCK:* Myosin light chain kinase · *MLCP:* Myosin light chain phosphatase · *MLC$_{20}$:* Myosin light chain 20 · *MP:* Myosin phosphatase · *MYPT1:* Targeting subunit of myosin phosphatase · *NCX:* Na/Ca exchanger · *NE:* Norepinephrine · *p160ROCK:* A rho kinase · *PAK:* P21-activated kinase · *PE:* Phenylephrine · *PGF2α:* Prostaglandin factor 2α · *PKC:* Protein kinase C · *PKC-α:* Protein kinase C-α · *PKN:* Rho effector, protein kinase C-related kinase · *PL:* Plasmalemma · *PLC:* Phospholipase C · *PL-jSR:* Plasmalemma-junctional sarcoplasmic reticulum · *PMA:* Phorbol 12-myristate 13-acetate · *PP1c:* Catalytic subunit of myosin phosphatase · *PSF:* Point spread function · *PMCA:* Plasmalemma Ca^{2+} pumping ATPase · *PM-SR:* Plasma membrane-sarcoplasmic reticulum · *ROK:* Rho-associated kinase · *RYR:* Ryanodine receptor · *SBB:* Superficial buffer barrier · *SERCA:* Sarcoplasmic reticulum Ca^{2+} ATPase · *Ser/Thr:* Serine/threonine · *SMC:* Smooth muscle cell · *SMPP-1M:* Smooth muscle phosphatase-1M · *SOC:* Store-operated channels · *SR:* Sarcoplasmic reticulum · *STOCs:* Spontaneous transient outward currents · *TnI:* Inhibitory subunit troponin I · *TPEN:* N,N,N′N′-tetrakis (2-pyridylmethyl) ethylenediamine · *Tyr:* Tyrosine · *UTP:* Uridine 5′-triphosphate · *VSMC:* Vascular smooth muscle cells · *ZIP kinase:* Zipper interacting protein kinase

Electronic Supplementary Material Supplementary material is available for this article if you access the article at http://dx.doi.org/10.1007/s10254-003-0019-8. A link in the frame on the left on that page takes you directly to the supplementary material.

Preview

Our intention in this review is to derive an updated, comprehensive hypothesis on the mechanisms of force generation and signal transduction during α_1-adrenoceptor-mediated activation of mammalian arterial smooth muscle. We will focus particularly on small arteries, as these strongly influence total peripheral resistance. Of course, changes in the concentration of intracellular calcium ions ($[Ca^{2+}]_i$) have a major role in adrenergic signaling, as the primary activator of contraction in smooth muscle is the Ca^{2+}-calmodulin complex. In this regard, the relatively recent application of confocal Ca^{2+} imaging in intact arteries has revealed a rich new set of subcellular Ca^{2+} signals: waves, oscillations, sparks, microsparks, flashes, and ripples. Most of these data were unavailable at the time of the last major reviews of Ca^{2+} in smooth muscle function (Horowitz et al. 1996; Karaki et al. 1997; however, see Sanders 2001; Lee et al. 2002a). Similarly, recent experiments on permeabilized arteries and single isolated cells have revealed the presence and characteristics of biochemical systems that change the relationship between $[Ca^{2+}]_i$ and contraction (viz. Ca^{2+}-sensitization or Ca^{2+}-desensitization). Yet, the contributions and interactions of the different protein kinases (e.g., protein kinase C, *rho*-associated kinase, mitogen-activated protein kinases) in influencing Ca^{2+} sensitivity during normal adrenergically activated contractions of arteries in situ are largely unknown. Finally, certain thin filament-associated proteins [CaD, calponin (CaP)] may modulate crossbridge cycling during adrenergic activation, and crossbridges may cycle slowly (the "latch" state). These mechanisms also influence apparent "Ca^{2+} sensitivity." Ultimately, of course, all the mechanisms activated by adrenergic stimulation that influence contraction must converge on the muscle crossbridges, formed by the interaction of thick (myosin) and thin (actin) filaments. It seemed natural, therefore, also to divide adrenergic signaling into the two broad categories of "thick filament regulation" and "thin filament regulation" (c.f. Morgan and Gangopadhyay 2001). Thus, we have focused in this review on describing Ca^{2+} signaling and associated events, and on the mechanisms of thick and thin filament regulation. Of course, the goal of providing a quantitative description of all these processes during adrenergic activation will certainly not be reached for some time yet. Nevertheless, our goal here is to provide a unified accounting, with respect to Ca^{2+} and the biochemical systems, of what is presently known with respect to α_1-adrenoceptor-mediated regulation of crossbridge cycling and hence, changes in arterial diameter.

Overviews

α_1-Adrenergic activation

Physiologically, arterial diameter is regulated in large part by the sympathetic nervous system, in which the major α_1-adrenoceptor agonist, norepinephrine (NE) is released from sympathetic nerves within the artery wall. Most experimental studies of α_1-adrenergic activation of resistance arteries, however, have utilized application of exogenous (i.e., bath-applied) adrenergic agonists, such as NE or the synthetic compound, phenylephrine (PE). In this regard, bath application of a relatively specific α_1-receptor agonist, such as PE, certainly does not mimic the release of NE, ATP, and vasoactive peptides at specialized sympathetic neuro-effector junctions (reviewed in Hirst and Edwards 1989; Mulvany and Aalkjaer 1990). Furthermore, as the type and distribution of receptors and innervation varies with species and vascular beds, it may be expected that the physiological relevance of bath-applied α_1-adrenergic agonists will also vary. Nevertheless, the maximal achievable isotonic contraction of rat mesenteric small arteries has been found to be the same for neuronal stimulation and for exogenously applied NE (Nilsson et al. 1986). In rat mesenteric small arteries, an initial small transient component of the contraction has been attributed to activation of purinergic (P2X) receptors (see Gitterman and Evans 2001 for recent work), and the contribution of this component is greater in small arteries than in larger ones. Furthermore, stimulation of sympathetic perivascular nerves produces localized postjunctional Ca^{2+} transients (termed jCaTs; Lamont and Wier 2002) in arterial smooth muscle cells. Such Ca^{2+} transients, which certainly are a component of contractile activation by the sympathetic nervous system, are not observed during bath application of α_1-adrenergic agonists. Thus, bath-application of specific α_1-adrenoceptor agonists, such as PE, to isolated rat mesenteric arteries provides a convenient method to activate α_1-adrenergic receptors, but does not mimic entirely the physiological activation of contraction by sympathetic neuronal activity.

The major biochemical events occurring after α_1-receptor occupation in vascular smooth muscle are well known (for recent reviews, see Somlyo and Somlyo 2000; Zhong and Minneman 1999) and will be reiterated here only briefly. Phospholipase C (PLC) is activated by the G protein Gα,q to hydrolyze phosphatidylinositol-bis-phosphate to inositol 1,4,5-trisphosphate (InsP$_3$) and diacylglycerol (DAG), which activates PKC. Ca^{2+} signaling is initiated directly by InsP$_3$, as Ca^{2+} is released from the sarcoplasmic reticulum (SR) subsequent to binding of InsP$_3$ to its receptors (SR Ca^{2+} release channels). Protein kinase signaling cascades, which may ultimately also influence crossbridge cycling, are initiated by the activation of conventional PKCs (viz. those that require both Ca^{2+} and DAG, such as α, β) and novel PKCs (viz. those that require only DAG, such as ε). PKCs may also be upstream regulators of numerous kinases, including MLCK, ERK1/2, rho-kinase (p160ROCK), and CaMKinase II, as well as various ion channels and ion transporters. In addition, rho-kinase may be activated more directly, with effects on crossbridge cycling, through a mechanism involving Gα,q and the nucleotide exchange factor (GEF), and CaMKinase II may be activated directly by the Ca^{2+} increase.

Force generation in smooth muscle is the result of crossbridge cycling, and for smooth muscle crossbridges to begin to cycle, the regulatory light chain of smooth muscle myosin (MLC$_{20}$) must be phosphorylated. The major mechanism of MLC$_{20}$ phosphorylation is MLCK, which is activated by Ca^{2+}-calmodulin. Dephosphorylation of MLC$_{20}$-P is mediated by smooth muscle myosin phosphatase (MP, MLCP, or SMPP-1 M). Dephosphorylated crossbridges may cycle slowly (i.e., the "latch" state). The activity of MLCP is regulated by signaling cascades involving both PKC and Rho-kinase. "Ca^{2+} sensitization of the myofilaments" can result from decreased activity of MLCP and therefore, increased MLC$_{20}$-P, for a given [Ca^{2+}]$_i$. Finally, "Ca^{2+} sensitization" may also be produced by removal of inhibitory influences on crossbridge cycling. This is mediated by thin filament-associated proteins (CaD) that are regulated by signaling cascades perhaps involving PKC and the MAP kinases, ERK1/2.

Selected aspects of arterial smooth muscle ultrastructure

As stated just above, our intention in this review is to focus on the phenomenology of adrenergically stimulated Ca^{2+} signaling and other mechanisms controlling contraction. This necessitates a brief review of the ultrastructure of vascular smooth muscle, with particular emphasis on the SR and the caveolae of the surface membrane (plasmalemma, PL). We show later in this review (in the section entitled "Revision") that Ca^{2+} signals thought previously to be directly related to Ca^{2+} influx through the plasmalemma in fact arise from cyclical release of Ca^{2+} from the SR. Similarly, an important role of caveolae in adrenergic signaling is now indicated by the fact that rho-A, rho-associated kinase (ROK), and PKC-α, all putatively involved in Ca^{2+} sensitization of contraction are translocated to the membrane after adrenergic activation (Taggart et al. 1999), and this translocation is inhibited by the presence of caveolin-1 scaffolding domain peptide (Taggart et al. 2000), indicating the importance of caveolae.

Caveolae

Caveolae are enriched in receptors, ion channels (e.g., Ca^{2+} channels or DHP receptors), signal transducers (e.g., Gα, Gβ, DAG, PKC-α, calmodulin), and effector molecules, such as adenylyl cyclase and MAP kinase (for review, see Shaul and Anderson 1998). In general, caveolae in vascular smooth muscle may be sites for the integration of events linking extracellular stimuli and intracellular effectors (see Taggart 2001). In fact, it has been suggested (Taggart 2001) that "receptor-coupled stimulation of smooth muscle involves both an increase in $[Ca^{2+}]_i$ and plasma membranous recruitment (to caveolae) of signaling molecules important for Ca^{2+}-sensitization of force production." Thus, the translocation of PKCε and MAP kinase to the membrane (Khalil and Morgan 1993) may involve caveolae, but the images lacked the resolution to show this definitively. The translocation of *rho*-A to the membrane after agonist stimulation has also been shown with biochemical binding studies (Gong et al. 1997), and it seems likely that caveolae are involved. Finally, caveolae may play a particularly important role in the generation of Ca^{2+} sparks, as they are known to contain L-type Ca^{2+} channels and exist in close juxtaposition to the SR, which contains the RyR. Disruption of caveolae by methyl-β-cyclodextrin (dextrin), which depletes the caveolar membrane of cholesterol, reduced the frequency of Ca^{2+} sparks in rat arterial smooth muscle cells (Shaul and Anderson 1998; Lohn et al. 2000). Caveolae are known to be enriched in β-adrenergic receptors, but the presence of α-adrenergic receptors in caveolae has not, to our knowledge, been demonstrated.

Sarcoplasmic reticulum

The sarcoplasmic reticulum (SR) of mammalian vascular smooth muscle comprises an extensive network throughout the cytoplasm, also enveloping the nucleus and mitochondria (Lesh et al. 1998). In rabbit portal vein myocytes, the superficial SR appears to be arranged spirally (Gordienko et al. 2001). The superficial SR makes surface coupling with the plasmalemma, and may be penetrated by the caveolae. Areas of very close apposition ($\approx$15 nm) between SR and PL (Somlyo and Franzini-Armstrong 1985) may provide the places for interactions between PL entities, such as ion channels and transporters, and corresponding SR entities, as well as providing a space in which diffusion of ions or other substances may be slowed. SR deep in the smooth muscle cell, on the other hand, will be in less direct communication with the surface membrane. In general, it is now recognized that the SR plays an exceedingly complex role in regulation of smooth muscle contraction, able to mediate both relaxation (through Ca^{2+} sparks and Ca^{2+} uptake; Nelson et al. 1995; Lee et al. 2002a), and contraction (through agonist-induced release of Ca^{2+}). Furthermore, the SR may be able to intercept Ca^{2+} entering the cell via the PL, a salient feature of the superficial buffer barrier (SBB) hypothesis (see Van Breemen et al. 1995). For purposes of this overview, several functions of the SR may be noted as being particularly relevant to adrenergic signaling: (a) the release of Ca^{2+} through RyR, which activates large conductance K^+ (BK) channels, (b) the release of Ca^{2+} through InsP$_3$R, which generates Ca^{2+} waves and may also, under certain circumstances, short circuit the SR, allowing Ca^{2+} entering the cell to penetrate past the SR into the myoplasm, (c) communication with store-operated channels (SOC) through which Na^+ (Arnon et al. 2000) and/or Ca^{2+} may enter when the SR becomes depleted of Ca^{2+}, and (d) the uptake of Ca^{2+} from the cytoplasm, permitting relaxation and store refilling. Indeed, the SR is proposed to "deliver" Ca^{2+} to the myofilaments, where calmodulin (CaM) is bound (Wilson et al. 2002). It has also been proposed that there are specialized junctions between plasmalemma and junctional regions of the SR, PL-jSR junctions, which together constitute a functional unit, the PlasmaERo-some (see Blaustein et al. 2002). Numerous molecular interactions occur at junctions between the plasma membrane and the SR, including an interaction between the InsP$_3$ receptor and Trp proteins, which may mediate store-operated Ca^{2+} entry, now increasingly being recognized as important in smooth muscle function (see Beech 2002), including the response to adrenergic agonists (Lee et al. 2002b; Zhang et al. 2002). We note that the role of the SR may be investigated now more definitively than in the past, as techniques have been developed to monitor SR [Ca^{2+}] levels (Shmigol et al. 2001; Golovina and Blaustein 1997).

The SR of vascular smooth muscle contains both InsP$_3$ receptors (three isoforms), and ryanodine receptors (three isoforms: RyR1, RyR2, RyR3). These release channels may exist in several different locations in the SR, and may release Ca^{2+} from functionally separate stores within the SR. Thus, through selective activation of spatially localized SR Ca^{2+} release, different functions, such as both contraction and relaxation, may be served. Immunohistochemical localization of RyR in aortic cells reveals RyR in both central and peripheral SR (Lesh et al. 1998). As noted by Iino (Iino 2002), Ca^{2+} is the activator of both RyR and InsP$_3$ receptors, thus creating the possibility of channel crosstalk (Gordienko and Bolton 2002). Functional evidence, such as gained from studies of Ca^{2+} release, has been used to suggest almost every possible different arrangement of RyR and InsP$_3$R on the SR: (a) a single store, containing both InsP$_3$R and RyR, (b) two separate stores, each containing only RyR or only InsP$_3$R, (c) two separate stores, one containing both, the other only InsP$_3$R, and (d) two stores, one containing both, the other only RyR (for a recent summary of these data, see Flynn et al. 2001). Even within the arterial smooth muscle of the same animal (dog), the functional evidence favors different arrangements in different arteries: Acutely isolated single canine pulmonary artery smooth muscle cells are thought to possess stores of type 3 above (RyR and InsP$_3$R on completely separate stores), while canine renal artery myocytes are thought to possess a store of type 1 above (both RyR and InsP$_3$R on the same store; Janiak et al. 2001). Cultured rat aortic cells are thought to have two separate Ca^{2+} stores (Tribe et al. 1994), but the arrangement of the SR of smooth muscle cells in culture is markedly different from that in intact arteries, perhaps changing as the cells move from the contractile phenotype to the proliferative phenotype. In both acutely isolated rat mesenteric artery myocytes, however (Baro and Eisner 1995), and in intact rat mesenteric arteries (Zang et al. 2001; C. Lamont and W.G. Wier, unpublished data), the evidence favors two functionally separate Ca^{2+} stores, one containing both RyR and InsP$_3$R, and one containing only RyR. Recently, an elegant and conclusive study of the Ca^{2+} stores in at least one other type of smooth muscle, colonic, reached a similar conclusion (Flynn et al. 2001).

Calcium signaling

Revision

Our conception of Ca^{2+} signaling in VSMC during adrenergic activation has undergone revision recently because of confocal imaging of [Ca^{2+}]$_i$ within single smooth muscle cells in the arterial wall (Iino et al.1994; Kasai et al.1997; Miriel et al. 1999; Ruehlmann et al.

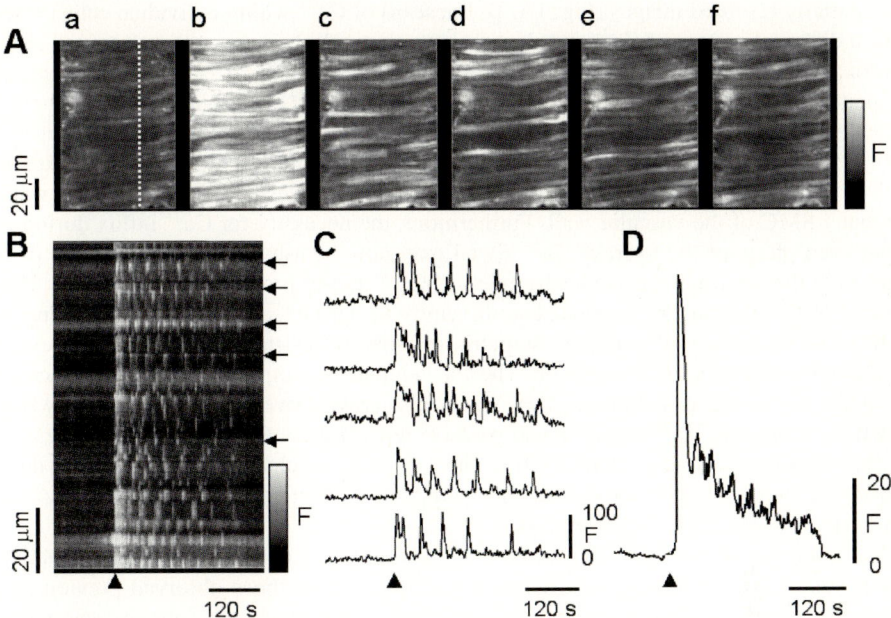

Fig. 1A–D Spatio-temporal changes in [Ca^{2+}] elicited by a high concentration of the α_1-adrenoceptor agonist, PE, in the individual smooth muscle cells of a rat small artery. **A** Images of fluo-4 fluorescence before (*a*), during (*b–e*), and after (*f*) exposure to PE (5.0 μM). The full sequence of 480 images is also available as a video clip (see video file), in the supplementary material of this publication and in the original publication (Zang et al. 2001). Image *b* shows the initial, brief, homogeneous rise in Ca^{2+} occurring within seconds of exposure to PE. Images *c–f* show the asynchronous Ca^{2+} waves within individual SMC. **B** "Virtual" line-scan image constructed from the same sequence of 480 images. *Upward arrowheads* in this figure indicate the time of application of PE. *Dotted line* in *a* (**A**) gives the position of the virtual scan-line. Ca^{2+} waves within individual SMC are evident as the periodic increases in fluorescence within each cell. Note that the Ca^{2+} waves within neighboring cells do not appear to be correlated (i.e., are asynchronous), and that the frequency of the Ca^{2+} waves declines with time during the maintained presence of PE. **C** Line-plots of fluo-4 fluorescence of individual cells obtained from the virtual line-scan image at the places indicated by the *arrows*. **D** The single trace represents the time course of the spatial average [Ca^{2+}] during the sequence of 480 images. The trace is the average fluo-4 fluorescence within the entire image. This is representative of the recording of fluo-4 fluorescence that would be obtained without any spatial resolution, i.e., as if fluo-4 fluorescence had been recorded without imaging. (From Zang et al. 2001)

2000; Mauban et al. 2001; Zang et al. 2001; see "Appendix" for a discussion of the methodology of imaging Ca^{2+} in arteries). The images in Fig. 1 (Zang et al. 2001) illustrate both the old view of Ca^{2+} signals, and the new. The trace in Fig. 1D represents the spatially averaged Ca^{2+}-dependent fluorescence from the sequence of confocal images of the arterial wall in Fig. 1A. It is similar, therefore, to what had been recorded previously, using Ca^{2+} indicators such as fura-2 and aequorin, neither of which normally provides a spatially resolved signal (although fura-2 can provide an image in thin cells). This type of signal (Fig. 1D) gave rise to the concept that [Ca^{2+}]$_i$ increases transiently and then declines to suprabasal levels. Many studies have shown that the initial transient increase in [Ca^{2+}]$_i$ arises from Ca^{2+} released from internal sites (SR), and that the later maintained elevation of [Ca^{2+}]$_i$ is dependent on external Ca^{2+} (as it is blocked or inhibited by agents that interfere with entry of Ca^{2+} through L-type Ca^{2+} channels, and through ROC or SOC channels).

The spatially resolved images (Fig. 1A, B, linescan) of Ca^{2+} within individual cells present quite a different picture of Ca^{2+} signals, however, particularly during the maintained elevation of spatially averaged $[Ca^{2+}]_i$. Within individual cells, $[Ca^{2+}]$ is actually undergoing the pattern of "baseline spiking," in which changes in $[Ca^{2+}]$ propagate as waves through the cells. Within individual cells, there is very little steady elevation of $[Ca^{2+}]_i$. The data in Fig. 1 make apparent the fact that the elevation of average $[Ca^{2+}]_i$ seen with indicators such as fura-2 and aequorin, is due to the summation of *asynchronous* Ca^{2+} waves within the many SMC of the vascular wall. Furthermore, the necessity for Ca^{2+} influx during the maintained phase of the average Ca^{2+} signal may now be ascribed to the necessity of replenishing the SR after it periodically releases Ca^{2+} during propagated Ca^{2+} waves. Thus, the role of the SR must be expanded significantly in our conception of Ca^{2+} signaling. At high levels of agonist stimulation, isometric or isobaric resistance arteries often develop oscillatory contraction or vasomotion. The Ca^{2+} signals accompanying this type of activity are distinct from the asynchronous Ca^{2+} waves discussed above (see "Adrenergically stimulated asynchronous Ca^{2+} waves"), and we have termed these Ca^{2+} transients *synchronous* Ca^{2+} oscillations (Mauban et al. 2001). Their cellular mechanism remains to be determined, although an hypothesis for the mechanism of their initiation has been advanced recently (Peng et al. 2001; Sell et al. 2002).

It is important to note also that the Ca^{2+} signals of individual VSMC within the walls of intact pressurized rat mesenteric arteries are different from those observed previously in enzymatically dissociated single rat mesenteric artery smooth muscle cells. Adrenergic activation of single isolated cells causes a single transient Ca^{2+} spike, which then declines to a level that is elevated above basal in the continued presence of the agonist (Baro and Eisner 1992, 1995; Baro et al. 1993). The reasons for the absence of repetitive Ca^{2+} waves in isolated cells are not known. While certain transformed smooth muscle cells (e.g., A7r5) do produce repetitive Ca^{2+} spikes (Blatter and Wier 1992), these cells are different from intact arterial cells in many ways.

In summary, the notion that $[Ca^{2+}]_i$ falls to steady, slightly elevated levels during maintained agonist action is not supported by the recent images of Ca^{2+} within single cells of intact arteries. Nevertheless, the efficacy of the asynchronous Ca^{2+} waves in activating contraction via phosphorylation of MLC_{20} is also not known. Thus, the relative contributions of Ca^{2+} activation and Ca^{2+} sensitization, particularly during tonic vasoconstriction, are not known. In the sections that follow, we review the available data on Ca^{2+} signaling during adrenergic activation of mammalian arteries. In general, this signaling involves asynchronous propagating Ca^{2+} waves, synchronous, spatially uniform Ca^{2+} oscillations, and Ca^{2+} sparks.

Adrenergically stimulated asynchronous Ca^{2+} waves

The first indication that adrenergic Ca^{2+} signaling in intact arteries can involve asynchronous Ca^{2+} waves, rather than spatially uniform changes in $[Ca^{2+}]_i$, came from the seminal study of Iino and his colleagues (Iino et al. 1994). These authors were the first to apply confocal microscopy to an intact preparation of mammalian artery. They used segments of rat tail artery which were stretched over flattened glass tubes. This helped immobilize the smooth muscle cells and provided an extended, flat preparation that could be placed close to the bottom of a recording chamber. Endothelium, smooth muscle cells, and perivascular nerves could all be visualized separately by changing the plane of focus (i.e.,

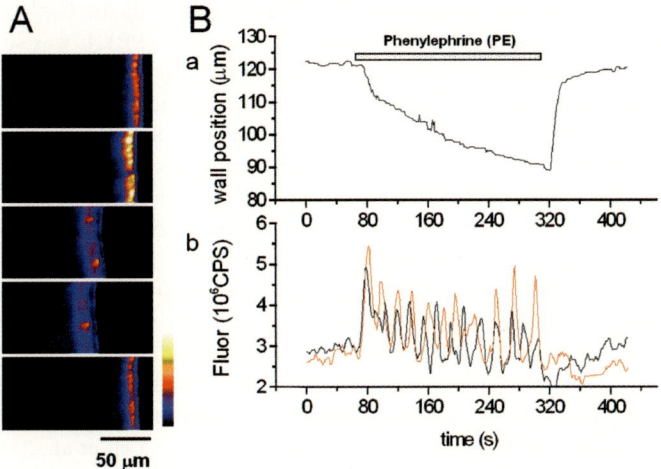

Fig. 2A, B Ca^{2+} and contraction recorded simultaneously during the stimulation of a pressurized (70 mmHg) rat mesenteric artery with PE (300 nM). **A** "Radial" view images of the arterial wall before (*top*), during (*middle frames*) and after removal of PE (*bottom*). SMC are seen in crosssection. Fluorescence intensity, indicating Ca^{2+}-dependent fluo-4 fluorescence, is represented with the color bar ([Ca^{2+}] increases with color from *bottom to top* of the color bar). **B** (*a*) Relative position of the arterial wall, as determined from images in **A**, in response to stimulation. The horizontal field-of-view is 150 μM, with the center of the artery to the *left*. **B** (*b*) *Red* and *black* traces in *b* (**B**) indicate fluorescence from two identified cells that could be followed throughout the experiment. Note that the fluorescence changes during the vasoconstriction are distinctly asynchronous amongst different cells. The full video sequence from which the images in **A** were obtained is available as supplementary data in the original publication. (Mauban et al. 2001)

the confocal optical section). The possibility of any small motion was eliminated through the use of cytochalasin D (5.0 μM). Stimulation of the perivascular nerves (e.g., 300 pulses at 5 Hz) evoked asynchronous Ca^{2+} waves that propagated within individual cells (velocity, ~20 μm/s). Application of noradrenaline in the bath (e.g., 3.0 μM) produced similar Ca^{2+} signals. Ryanodine (applied transiently and simultaneously with 50 mM caffeine) blocked the asynchronous Ca^{2+} waves. The frequency of the asynchronous Ca^{2+} waves increased with the concentration of bath-applied noradrenaline. Asynchronous Ca^{2+} waves elicited by adrenergic stimulation (Fig. 2) have now been reported again in immobilized rat tail arteries (Kasai et al. 1997; Asada et al. 1999; Zang et al. 2001), in isobaric rat mesenteric small arteries (Miriel et al.1999; Mauban et al. 2001; Peng et al. 2001) and in immobilized rat inferior vena cava (IVC; Ruehlmann et al. 2000; Lee et al. 2001). As discussed in more detail later (Appendix), we believe that the isobaric small artery (constant transmural pressure) is an experimental preparation in a nearly physiological condition. Thus, the experimental demonstration of asynchronous Ca^{2+} waves during adrenergically-stimulated decrease in diameter of a pressurized small artery at mammalian temperature (Fig. 2; Mauban et al. 2001) establishes, unequivocally, the physiological relevance of these Ca^{2+} signals.

Despite the unequivocal occurrence of Ca^{2+} waves during vasoconstriction, it is not yet completely clear to what extent the asynchronous Ca^{2+} waves are actually involved in activating contraction. For example, the possibility exists that such waves, because of their brevity, might be relatively less effective, without a background of "Ca^{2+} sensitization." A spatially heterogenous pattern of contractile activation within the arterial wall might be less effective than a spatially uniform pattern. The frequency of the Ca^{2+} waves does increase, however, as a function of the concentration of agonist (PE). It was suggested, after their first observation (Iino et al. 1994), that "a graded response to different levels of the sympathetic transmitter (Bao and Stjarne 1993) seems to be accomplished not by a graded response within each muscle cell, but by a graded number of active cells within the vascular wall." This observation has been confirmed in a more recent and detailed study, also using immobilized rat mesenteric arteries; increasing concentrations of PE (0.1–10.0 μM) were associated with increasing numbers of cells that produce asynchronous Ca^{2+} waves and increasing frequency of Ca^{2+} waves (Zang et al. 2001). Strong support for a role of asynchronous Ca^{2+} waves in activating contraction of vascular smooth muscle comes from a series of studies (Ruehlmann et al. 2000; Lee et al. 2001; Lee et al. 2002a) in which Ca^{2+} was imaged confocally in isometrically mounted, de-endothelialized, everted rings of rabbit IVC. Increasing concentrations of PE were associated with a graded contraction, and with a graded recruitment of cells to produce asynchronous Ca^{2+} waves, and an increase in the frequency of the waves (Ruehlmann et al. 2000; Lee et al. 2002a). Application of the putative blocker of store-operated channels, SKF96365, in the presence of PE and nifedipine, abolished both Ca^{2+} waves and contractile force. Thus, the data from large veins (IVC) indicate that asynchronous propagating Ca^{2+} waves play a role in activating contraction. Even so, it has been reported recently that Ca^{2+} waves do not occur during the tonic phase of agonist-induced contraction of human saphenous vein (Crowley et al. 2002). In that tissue, tonic contraction seems to be maintained by an increase in Ca^{2+} sensitivity, mediated by *rho* kinase and tyrosine phosphorylation. On an apparently contrary note, it has been concluded recently that the net effect of asynchronous Ca^{2+} waves (and Ca^{2+}sparks) induced by pressure in rat cerebral arteries, is "to oppose contraction" (Jaggar 2001; note that this statement applies to pressure-induced Ca^{2+} waves, not agonist-induced waves; such Ca^{2+} waves could have a different mechanism and effect).

Mechanisms of agonist-stimulated asynchronous Ca^{2+} waves

Ca^{2+} release: RyR and InsP$_3$R

The fact that adrenergic agonists induce asynchronous propagating Ca^{2+} waves in small arteries is now well established. A major question remaining, however, is the identity of the SR Ca^{2+} release channels that are involved. The waves could, in theory, propagate via InsP$_3$-induced Ca^{2+} release (IICR), involving InsP$_3$ receptors, or via Ca^{2+}-induced Ca^{2+} release (CICR), involving ryanodine receptors, or a combination of both (Blatter and Wier 1992). It was proposed originally (Iino et al. 1994) that the asynchronous Ca^{2+} waves propagate mainly, or even exclusively, via sequential InsP$_3$-induced Ca^{2+} release (Iino et al. 1994). According to this mechanism, the agonist elevates the concentration of InsP$_3$ throughout the cell, lowering the threshold for Ca^{2+} activation of release through the InsP$_3$ receptor. As Ca^{2+} is released, first from some particular location, some of it diffuses onward to neighboring InsP$_3$ receptors, where it induces further release, with the result that a change of Ca^{2+} propagates through the cell (i.e., a Ca^{2+} wave). In support of this type of mechanism, interesting new evidence is that arteries lacking functional ryanodine receptors, as judged by the lack of response to caffeine and an absence of Ca^{2+} sparks, produce Ca^{2+} waves in response to noradrenaline that are identical to those produced by normal arteries (Dreja et al. 2001). (Arteries in this condition were produced by chronic exposure to

ryanodine; the means by which this produces nonfunctional ryanodine receptors is unknown.) This result implies that RyR are not necessarily involved in the asynchronous
propagating Ca^{2+} waves. On the other hand, recent evidence for the involvement of RyR
in Ca^{2+} waves of rat mesenteric arteries (Peng et al. 2001) was that ryanodine abolished
both asynchronous Ca^{2+} waves and synchronous Ca^{2+} oscillations. Interpretation of this result should consider the fact that ryanodine is capable of depleting all Ca^{2+} stores in vascular smooth muscle (i.e., those releasable by caffeine, $InsP_3$, and the Ca^{2+} ionophore,
A23187; Hwang and Van Breemen 1987; Kanmura et al. 1988). We note that interpretation of the effects of ryanodine is problematic, because previous reports on the effects of
ryanodine are apparently contradictory, including: (a) an increase in the contraction and
an abolition of vasomotion (Gustafsson et al. 1994), (b) a promotion of vasomotion where
it had not occurred before (Omote et al. 1993), (c) little affect other than on the rate at
which the calcium and force levels were attained (Garcha and Hughes 1995; Julou-Schaeffer and Freslon 1988), and (d) intermediate responses (Ashida et al. 1988; Boittin et al.
1999). In summary, the possible involvement of RyR in Ca^{2+} waves in arterial muscle remains unclear. In venous myocytes, on the other hand, rather convincing evidence was obtained that spontaneous Ca^{2+} release (due to basal activity of phospholipase C, PLC) and
spontaneous propagating Ca^{2+} waves in those cells involves "loose coupling" and "crosstalk" between RyR and $InsP_3$ receptors (Gordienko and Bolton 2002). Direct evidence on
the involvement of $InsP_3$ receptors derives mainly from the use of putative blockers of the
$InsP_3$-receptor, such as 2-aminoethoxydiphenylborate (2-APB) and Xestospongin C (Gordienko and Bolton 2002). 2-APB abolishes asynchronous Ca^{2+} waves in rabbit IVC (Lee
et al. 2002b) and in pressurized rat mesenteric small arteries (C. Lamont and W.G. Wier,
unpublished data). Similar to the issue with ryanodine, however, 2-APB may also deplete
Ca^{2+} stores, by blocking Ca^{2+} entry through store-operated-channels (SOCs; see "Ca^{2+} influx: ROCs, SOCs, VOCs, and NCX").

Recently, a detailed hypothesis on the mechanism of agonist-induced asynchronous
propagating Ca^{2+} waves has been described by Van Breemen and his colleagues (Lee et
al. 2002a). Particularly noteworthy is their suggestion that Ca^{2+} is released through $InsP_3$
receptors located mainly in "radial" SR, near CaM that is tethered to the myofilaments. In
this scheme, Ca^{2+} entering the cell is prevented from occupying a "myosin-poor" space
near the membrane, because it is taken up by SERCA and delivered by radial SR (during a
Ca^{2+} wave) to the "myosin-rich" space in the interior of the cell. The authors note that this
scheme can provide an explanation for the observation that inhibition of SERCA changes
the relationship between spatial average $[Ca^{2+}]$ and force by allowing Ca^{2+} to occupy the
"myosin-poor" space. It may be possible to verify this hypothesis directly, by confocal imaging Ca^{2+}.

Ca^{2+}influx: ROCs, SOCs, VOCs, and NCX

Whether or not the release of Ca^{2+} involves RyR in addition to the InsP3 receptors, Ca^{2+}
influx will be required after a wave, because a fraction of the Ca^{2+} that is released is then
extruded from the cell, via the forward mode of the Na/Ca exchanger (NCX) and via the
plasmalemma Ca^{2+} pumping ATPase (PMCA). In intact arteries, Ca^{2+} stores are reported
to refill, in the maintained presence of agonist, by at least three mechanisms (Lagaud et al.
1999): (a) the Na/Ca exchanger, (b) channels that are insensitive to nitrendipine, but are
blocked by SKF 96365 (SOCs), and (c) nitrendipine-sensitive channels (presumably volt-

age-dependent L-type Ca^{2+} channels) that are activated by tyrosine kinase. Voltage-dependent Ca^{2+} channels may also be activated by depolarization and contribute Ca^{2+} influx for refilling the SR. Recent evidence points to a surprisingly important role for SOCs, however, in maintaining adrenergically-activated contraction, as described next.

In rat mesenteric small arteries, elevation to 10.0 mM of external Mg^{2+}, blocks SOCs (Hoth and Penner 1993), and rapidly abolishes PE-induced Ca^{2+} waves (Zhang et al. 2002) during the tonic phase of the agonist-induced contraction. Similarly, agonist-induced Ca^{2+} waves in rabbit inferior vena cavae are abolished by 2-APB (Lee et al. 2002b), a blocker of SOCs (Diver et al. 2001) and $InsP_3$ receptors. Interpretation of these results is problematic because the relative efficacy of 2-APB in blocking SOCs and $InsP_3R$ is unknown and because 2-APB may also block SERCA directly (Missiaen et al. 2001). The existence and operation of SOCs in rat mesenteric small arteries may be demonstrated by store-depletion protocols. Readmission of Ca^{2+} to an artery depleted of its Ca^{2+} stores by exposure to the Ca^{2+} chelating compound, TPEN, is accompanied by a transient contraction (Zhang et al. 2002). This is conventionally attributed to entry of Ca^{2+} through SOCs, as a result of activation of SOCs by store depletion.

Electrophysiological studies show that adrenergic stimulation (noradrenaline) activates a nonselective cation current in rabbit portal vein myocytes (Aromolaran et al. 2000), and evidence was obtained that the channel conducts Ca^{2+}, and is a SOC (Albert and Large 2002). Nevertheless, studies in excised patches showed that this channel can also be activated independently of store depletion, by PKC activators. The fact that a channel may be activated by store depletion, as well as by receptor-associated mechanisms, blurs the distinction between an SOC and a ROC (see Beech 2002 for a recent "perspective" on these results).

The mechanisms of Ca^{2+} entry after store depletion, however, do not seem to be simply Ca^{2+} entry directly through SOCs or ROCs. Rather, Ca^{2+} entry is thought to occur through the Na/Ca exchanger and through voltage-dependent Ca^{2+} channels, secondarily to entry of Na^{2+} through nonselective SOCs. In this case, accumulation of Na^{2+} in the sub-PL space would activate the NCX in reverse mode (Arnon et al. 2000), and depolarization resulting from Na^+ and Ca^{2+} through SOCs would activate voltage-dependent Ca^{2+} channels. Indeed, it has been proposed that the basis of α_1-adrenergic activation of venous smooth muscle is the initial opening of $InsP_3$ receptors of the SR (resulting in store depletion), followed by opening of SOCs and inward current, depolarization, and further Ca^{2+} entry through NCX and voltage-dependent Ca^{2+} channels (Lee et al. 2002b).

Surprisingly, nifedipine completely blocked the contraction (Zhang et al. 2002) that follows readmission of external Ca^{2+} after store depletion. Within the context of the information given above, two possible explanations are: (a) nifedipine blocks SOCs, (b) nifedipine blocks voltage-gated Ca^{2+} channels that are activated secondarily by depolarization that results from inward current through SOCs. Explanation (b) implies either that Ca^{2+} entry through SOCs is negligible (current would be carried by Na^+ instead), or that the Ca^{2+} that does enter through SOCs, does not become available for contraction. [It has been proposed that the Ca^{2+} that enters through SOCs of rabbit arteriolar smooth muscle enters into a "noncontractile" compartment (Flemming et al. 2002), perhaps the "myosin-poor" space mentioned (see "Ca^{2+} release: RyR and $InsP_3R$").] Although the dihydropyridines (DHPs) such as nifedipine are widely viewed as selective blockers of L-type Ca^{2+} channels (Fleckenstein-Grun 1996; Schwartz 1994; Van Zweiten and Pfallendorf 1993) and thus are potentially useful in distinguishing the role of L-type Ca^{2+} channels from other channels, there are reports that DHPs may block other Ca^{2+} entry channels including SOCs and/or ROCs (Cauvin et al. 1988; Curtis and Scholfield 2001; Stepien and Marche 2000; Zhang et al. 2002). These results raise the possibility that effects of nifedipine may be due to the blocking of SOC, as well as L-type Ca^{2+} channels.

The role of the Na/Ca exchanger in Ca^{2+} entry to support SR-dependent Ca^{2+} waves in IVC has also been demonstrated recently in Ca^{2+} imaging studies. Ca^{2+} waves in IVC pre-exposed to nifedipine and activated by PE (5.0 μM) were abolished by the NCX inhibitor, 2,4-dichlorobenzamil (2,4-DCB; Lee et al. 2002b). (In IVC, nifedipine alone reduces the frequency of agonist-induced waves to about half, compared to control.) This result indicates the essential role of NCX in refilling Ca^{2+} stores during agonist-induced contractions.

Ca^{2+} efflux and sarcoplasmic reticulum Ca^{2+} uptake: Ca^{2+} pumping ATPase, Na/Ca exchanger and SERCA

Two mechanisms are known to play a role in net Ca^{2+} extrusion from arterial smooth muscle, the ATP-driven plasma membrane Ca^{2+} pump (PMCA; Nelson et al. 1997; Wuytack et al. 1992) and the Na^+ electrochemical gradient-driven Na^+/Ca^{2+} exchanger (NCX; Blaustein and Lederer 1999). The role of the latter has been reviewed recently (Blaustein and Lederer 1999), and therefore need not be reiterated here. As might be expected, SERCA has a vital role in agonist-induced asynchronous propagating Ca^{2+} waves. Application of cyclopiazonic acid (CPA) to Ca^{2+} waves induced by PE (5.0 μM) in IVC results in abolition of the Ca^{2+} waves and elevation of Ca^{2+} (Lee et al. 2002b).

Synchronous Ca^{2+} oscillations

A salient feature of small arteries activated by high levels of bath-applied adrenergic agonists is that they often undergo rhythmic vasomotion. In situ, arterial vasomotion is thought to be important in increasing hydraulic conductance, without decreasing resistance (Parthimos et al. 1999). Mauban and colleagues showed recently (2001) that vasomotion in rat mesenteric small arteries is accompanied by $[Ca^{2+}]$ changes that are spatially uniform within the arterial wall (Fig. 3). These changes in $[Ca^{2+}]$ are thus quite distinct from the asynchronous propagating Ca^{2+} waves discussed above. We suggested that the cellular mechanisms of these changes are also quite distinct from those of the agonist-stimulated asynchronous Ca^{2+} waves, most likely involving oscillatory changes in membrane potential. It is quite clear that these changes in $[Ca^{2+}]$ are directly related to changes in arterial diameter, as oscillatory vasomotion is always accompanied by such Ca^{2+} oscillations. Spatially synchronous Ca^{2+} transients induced by adrenergic agonists have also been observed in isometrically mounted rat mesenteric arteries (Peng et al. 2001). Interestingly, such Ca^{2+} signals have not yet been reported for IVC. Vasomotion of pressurized arteries continues for the duration of agonist exposure during experimental recordings. Thus, in pressurized arteries, modulation of arterial diameter by Ca^{2+} continues indefinitely during α_1-agonist presence. Mechanisms that change Ca^{2+} sensitivity may be activated during this time as well; but, in pressurized arteries, oscillatory, spatially uniform changes in Ca^{2+} continue to modulate arterial tone indefinitely.

Recently, a hypothesis on the initiation of adrenergically stimulated vasomotion in isometrically mounted rat mesenteric small arteries was advanced (Peng et al. 2001). This hy-

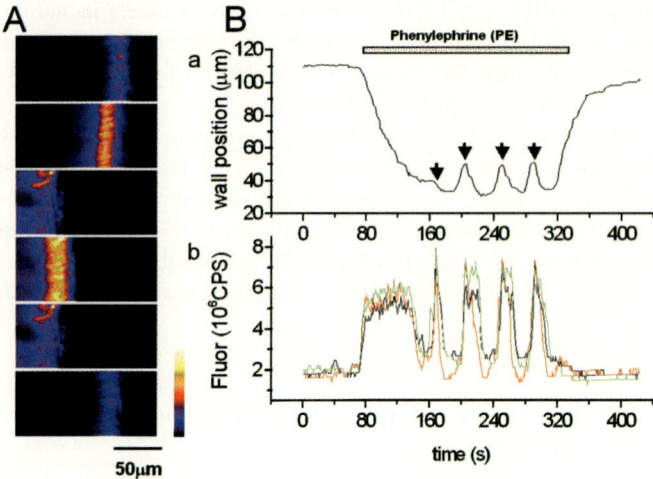

Fig. 3A, B Synchronous Ca^{2+}-transients and oscillatory vasomotion recorded simultaneously in a pressurized (70 mmHg) rat mesenteric artery stimulated with PE (1.0 μM). **A** A "Radial" view images of the arterial wall before (*top*), during (*middle frames*) and after removal (*bottom*) of PE. SMC are seen in crosssection. Images are of Ca^{2+}-dependent fluo-4 fluorescence, as represented with the color bar. **B** (*a*) Relative position of the arterial wall, as determined from images in **A**. The horizontal field-of-view is 150 μm, with the center of the artery to the *left*. The *arrowheads* indicate the peak of the synchronous Ca^{2+} transients in *b* (**B**), *below*. The peak of the Ca^{2+} transients coincided with the maximum diameter during the oscillatory vasomotion. **B** *b*: *Red, green* and *black* traces in *b* (**B**) indicate average fluorescence from three identified cells that could be followed throughout the experiment. The full video sequence from which the images in **A** were obtained is available as supplementary data in the original publication. (Mauban et al. 2001)

pothesis may be reiterated as follows: adrenergic stimulation first initiates asynchronous Ca^{2+} waves of the type described above. These Ca^{2+} waves activate an inward membrane current, which tends to depolarize cells. At some point, sufficient current is activated to result in depolarization of all the smooth muscle cells. Depolarization activates voltage-gated Ca^{2+} channels and the resulting (synchronous) Ca^{2+} entry synchronizes Ca^{2+} release in all the cells. The cycle then continues. In rat mesenteric small arteries, synchronized Ca^{2+} oscillations depended on an intact endothelium (Peng et al. 2001). De-endothelialized arteries could also develop synchronous Ca^{2+} oscillations if 8-bromo-cGMP was added to the bath. This raises the obvious possibility that it is endothelial nitric oxide (NO) which is essential for adrenergic vasomotion, at least in that tissue. A different approach to understanding the cellular mechanisms of vasomotion has been taken by Griffith and his colleagues (Parthimos et al. 1999). Here, arterial chaos of rabbit ear arteries (pressure fluctuations) was explained in a mathematical model in which the essential feature is the nonlinear interaction of intracellular and membrane oscillators that depend on cyclic release of Ca^{2+} from internal stores and cyclic influx of extracellular Ca^{2+}, respectively. The control variables were (a) cytosolic [Ca^{2+}], (b) [Ca^{2+}] in the ryanodine-sensitive SR, (c) membrane potential, and (d) the open-state probability of Ca^{2+}-activated K$^+$ channels. Of course, variables such as cytosolic [Ca^{2+}] were a function of several cellular processes; Ca^{2+} influx through receptor-operated channels, Ca^{2+} release from InsP$_3$-sensitive stores, Na$^+$/Ca^{2+} exchange, and others. Force was calculated from the total number of crossbridges, using the latch-state model (Hai and Murphy 1988). In general, this model is a "common pool" model, in the sense that the spatial relationships of ion channels, SR, and plasma membrane are not considered. In addition, separate ryanodine-sensitive and InsP$_3$-sensitive Ca^{2+} stores were assumed. This model is, in general, quite successful in reproducing a variety of experimentally observed phenomena, including the effects of ryanodine and certain ion-channel blockers. Nevertheless, further development of models of intracellular Ca^{2+} signaling during vasomotion should include the ability to produce spatially localized Ca^{2+} signals, such as Ca^{2+} waves, since Ca^{2+} waves may be a precursor to the synchronous Ca^{2+} oscillations.

Ca^{2+} sparks and microsparks

Ca^{2+} sparks in smooth muscle were first recorded in isolated myocytes (Nelson et al. 1995). Interestingly, Ca^{2+} sparks were not seen in the first confocal imaging studies of intact arterial smooth muscle (Iino et al. 1994), probably because of the use of an objective lens with relatively low numerical aperture in that study, which limited spatial resolution. The existence of Ca^{2+} sparks in the walls of intact small arteries under physiological conditions of intraluminal pressure was confirmed in the first confocal imaging study of pressurized small arteries (Miriel et al. 1999), where a high resolution (N.A., 1.4) objective lens was used. An emerging concept for the role of Ca^{2+} sparks in vascular smooth muscle function is that voltage-dependent Ca^{2+} channels, ryanodine receptors, and Ca^{2+}-activated K$^+$ (BK) channels form a functional unit for regulating arterial tone (Jaggar et al. 1998b). Ca^{2+} entering the PM-SR junctional space via voltage-dependent L-type Ca^{2+} channels activates the ryanodine receptors to release Ca^{2+} producing a Ca^{2+} spark. As recently reviewed in detail by others (Jaggar et al. 2000), this Ca^{2+} then activates the BK channels to conduct K$^+$ outward (spontaneous transient outward currents, or STOCS; Benham and Bolton 1986), which tends to hyperpolarize the membrane and reduce Ca^{2+} entry via the voltage-dependent Ca^{2+} channels. The importance of Ca^{2+} sparks in regulating arterial tone in intact animals is indicated by the fact that animals in which Ca^{2+} sparks have been rendered ineffective (by knocking out the β1 subunit of the BK channel) have higher than normal blood pressure (Brenner et al. 2000). This elegant study provides direct functional evidence of the vasodilatory role of Ca^{2+} sparks in arterial smooth muscle.

Origins of Ca^{2+} sparks in vascular smooth muscle

There is little doubt that arterial muscle contains mRNA for RyR1, RyR2, and RyR3 and that Ca^{2+} sparks arise from Ca^{2+} released through ryanodine receptors (Lohn et al. 2001). Using antisense techniques in cultured portal venous cells, Coussin et al. (2000) have shown that RyR1 and RyR2 are both required for Ca^{2+} sparks, but deletion of RyR$_3$ had no effect. In contrast, a role for RyR$_3$ in smooth muscle Ca^{2+} sparks has been suggested recently (Lohn et al. 2001; Knot et al. 2001). The experimental evidence from cells and cerebral arteries of RyR3-deficient mice was that RyR3 normally inhibits Ca^{2+} spark production; the frequency of spontaneous Ca^{2+} sparks was approximately twice as high in cells from RyR3$^{-/-}$ animals as in cells from wild-type animals. In both isolated single guinea-pig mesenteric arterial smooth muscle cells (Pucovsky et al. 2002) and in pressurized rat mesenteric arteries (Miriel et al. 1999), Ca^{2+} sparks arise preferentially and repetitively at only certain sites, termed frequent discharge sites, or FDS (Gordienko et al. 1998). In isolated rabbit portal vein myocytes, the FDS have been shown to be located near the superficial SR network, usually within 1–2 μm of the surface membrane (Gordienko et al. 2001). Recently, Ca^{2+} release events smaller in duration, spatial spread, and amplitude than usual were recorded in isolated guinea-pig mesenteric myocytes (Pucovsky et al. 2002). These events were termed microsparks and it was suggested that they were underlying events to Ca^{2+} sparks. Thus, as in other tissues, Ca^{2+} sparks almost certainly arise from Ca^{2+} flux through multiple ryanodine receptors. The frequency of Ca^{2+} sparks is increased by depolarization (Jaggar et al. 1998a), suggesting that, as in cardiac muscle, Ca^{2+} sparks are triggered by the flow of Ca^{2+} through voltage-dependent channels (Lopez-Lopez et al. 1995).

Effects of adrenergic stimulation on Ca^{2+} sparks

Adrenergic activation (PE) decreases the frequency of Ca^{2+} sparks in rat mesenteric small arteries (Mauban et al. 2001) during the typical adrenergic vasoconstriction. Earlier studies had shown that uridine 5′-triphosphate (UTP), which binds to purinergic receptors (P2Y), also decreases Ca^{2+} spark frequency (Jaggar and Nelson 2000). Although UTP is not a physiological neurotransmitter, the ATP released during sympathetic nerve activity will bind to the P2Y receptors. It is not yet known, from experimental observation, how ATP might affect Ca^{2+} spark frequency. Activation of protein kinase C (PKC) by phorbol esters (e.g., PMA, phorbol 12-myristate 13-acetate) or DOG (1,2-dioctanoyl-*sn*-glycerol) decreased Ca^{2+} spark frequency in rat cerebral arteries (Bonev et al. 1997).

In single isolated guinea-pig mesenteric artery myocytes, application of noradrenaline increased the frequency of Ca^{2+} sparks and led to a Ca^{2+} wave (Pucovsky et al. 2002), followed by a period in which sparks were absent. The initial increase in frequency that led to a Ca^{2+} wave seems difficult to reconcile with the fact that adrenergic stimulation and activators of PKC both decrease spark frequency. It seems clear that RyR can participate in Ca^{2+} waves in vascular smooth muscle cells. Nevertheless, we suggested earlier that the mechanism of the asynchronous propagating Ca^{2+} waves in intact arteries is primarily release of Ca^{2+} through InsP$_3$R, rather than through RyR. In intact arteries, it seems likely that the depolarization rapidly induced by noradrenaline will activate voltage-gated Ca^{2+} channels, and lead to a transient increase in frequency of Ca^{2+} sparks. The effect of activated PKC to inhibit spark frequency may occur more slowly (Bonev et al. 1997). Indeed, the decline in Ca^{2+} spark frequency observed after PE application to intact mesenteric arteries occurred over 2 min (Mauban et al. 2001). If asynchronous propagating Ca^{2+} waves in intact arteries are generated mainly by Ca^{2+} released from the SR via InsP$_3$R, rather than through RyR, then adrenergic stimulation can increase the frequency of Ca^{2+} waves, while decreasing the frequency of Ca^{2+} sparks.

Ca^{2+} ripples and flashes

Recently, two other types of spontaneous Ca^{2+} signals, "ripples" and "flashes," have been observed, using wide-field fluorescence microscopy, in muscle cells of rat tail artery (Asada et al. 1999). Ca^{2+} ripples, which are much smaller in amplitude than Ca^{2+} waves elicited by noradrenaline, occurred spontaneously in about half of the cells. The velocity of propagation however, was similar to that of the larger, noradrenaline-induced Ca^{2+} waves. It was suggested that the Ca^{2+} ripples are generated via inositol 1,4,5-trisphosphate-induced

Ca^{2+} release in response to locally produced angiotensin II. It is possible that the frequency of ripples or proportion of SMC generating Ca^{2+} ripples and flashes could be affected by adrenergic stimulation. Another type of unique Ca^{2+} transient was first reported in that study, namely Ca^{2+} flashes. These were very rapid, large increases in Ca^{2+} that were initiated in a relatively small area (<20 μm) and spread in an (apparently) passive way. The flashes were observed so infrequently that they could not be studied reliably. The authors speculated that Ca^{2+} flashes were not due to locally activated voltage-dependent Ca^{2+} channels. It seems likely to us, however, that these signals must arise during spontaneous action potentials. Indeed, in pressurized rat mesenteric arteries in which neurotransmission is totally blocked, electrical stimulation of the smooth muscle cells elicits rapid, spatially uniform Ca^{2+} transients (C. Lamont and W.G. Wier, unpublished observations) similar in some ways to the spontaneous Ca^{2+} flashes reported in rat tail artery. Ca^{2+} flashes may occur physiologically, since spontaneous action potentials have been recorded from pressurized guinea-pig mesenteric small arteries (Zelcer and Sperelakis 1982).

Summary of Ca^{2+} signaling in vascular smooth muscle

Ca^{2+} sparks in cardiac and skeletal muscle are clearly elementary events of Ca^{2+} transients that activate contraction. They appear to play a fundamentally different role in vascular smooth muscle, however, namely that of signaling for relaxation rather than contraction. The Ca^{2+} that binds to CaM and does ultimately activate smooth muscle contraction (see next section) appears to be derived from the SR, and to be released into the cytoplasm through $InsP_3$-sensitive SR Ca^{2+} release channels ($InsP_3R$). In this scheme, $InsP_3$ receptors on the SR are used to supply the Ca^{2+} for contractile responses, and ryanodine receptors on the SR are used to supply local Ca^{2+} for activation of ion channels (viz. Ca^{2+} sparks and STOCs) that tend to oppose contraction (i.e., are vasodilatory). The speed of propagation of the $InsP_3R$-mediated Ca^{2+} waves is relatively slow (~20 μm s^{-1}), and should provide sufficient time for binding of Ca^{2+} to CaM. This spatio-temporal heterogeneity of $[Ca^{2+}]$ might be expected possibly to result in spatio-temporal heterogeneity of cell activation, but information on this point is entirely lacking. Arteries contract when asynchronous propagating Ca^{2+} waves are being generated; the higher the frequency of the waves and the greater the proportion of cells generating waves, the larger the contraction. Under high levels of adrenergic activation, the changes in $[Ca^{2+}]$ become spatially uniform, and may arise primarily from Ca^{2+} entering the cell via voltage-dependent Ca^{2+} channels during oscillations of membrane potential. These changes in $[Ca^{2+}]$ are clearly associated with changes in contraction (viz. vasomotion). We have focused so far on the changes in Ca^{2+} that might activate CaM (which activates MLCK and CaMKinase II). Nevertheless, changes in cytoplasmic $[Ca^{2+}]$ may also be expected to participate in the activation of certain other protein kinases (e.g., the conventional isoforms of PKC) and enzymes (e.g., phospholipase C), but little information is available now on how Ca^{2+} might be involved.

Calmodulin level and regulation

CaM general properties

Calmodulin (CaM) plays a key role as a Ca^{2+} binding protein in the regulation of vascular smooth muscle cells. A major function is the activation of MLCK, leading to phosphorylation of the myosin light chain (LC_{20}) and crossbridge cycling. Each molecule of CaM contains 4 E-F hand-type Ca^{2+}-binding domains, and upon binding Ca^{2+}, each domain takes on an open conformation that exposes hydrophobic residues and leads to the binding of target sequences. Thus, these conformational changes are thought to be essential in transducing changes in $[Ca^{2+}]_i$ into biochemical effects.

CaM is sensitive to the physiologic range of intracellular $[Ca^{2+}]$ ($K_d=5\times10^{-7}$–5×10^{-6}M; Chin and Means 2000). However, for several reasons, the precise affinity of CaM for Ca^{2+} is a moving target. Firstly, the C-terminal pair of E-F hands have a three- to fivefold higher affinity for Ca^{2+} than the N-terminal pair of sites (Chin and Means 2000). Secondly, the affinity for Ca^{2+} is altered by CaM binding to target proteins, sometimes in a dramatic fashion such as occurs with CaMKII (see "Role of CaMII in smooth muscle"). Thirdly, it is possible that posttranslational modification of CaM may regulate its interactions with both $[Ca^{2+}]_i$ and target proteins.

Regarding the possibility of posttranslational modification of CaM, phosphorylation of CaM on Ser/Thr has been reported in rat liver cells by casein kinase II (Quadroni et al. 1994). Phosphorylation of CaM on Tyr has also been reported in rat liver and adipocytes as well as in human placenta by insulin-receptor kinase (Sacks and McDonald 1989). Whether CaM is phosphorylated in differentiated smooth muscle is unknown. Past studies have shown that either Tyr- or Ser/Thr- phosphorylation of CaM is associated with decreased affinity for target proteins (Quadroni et al. 1994; Sacks and McDonald 1989).

Subcellular distribution of total CaM

The relative subcellular distribution of CaM in vascular smooth muscle cells is not known; however, in neural cells, it has been reported that activation by Ca influx causes a translocation of CaM from the cytoplasm to the nucleus (Deisseroth et al. 1998). Thus, the distribution of total CaM between subcellular compartments may well be dynamically regulated and could feasibly be altered by α-adrenergic agonists. Recently, it has been suggested that there is a pool of immobilized CaM, bound to thin filaments in a Ca^{2+} independent manner, similar to that served by troponin C (Wilson et al. 2002)

Availability of CaM

The total CaM content of smooth muscle is approximately 40 μM (Meisheri et al. 1985; Ruegg et al. 1983; Zimmermann et al. 1995). However, it is often assumed that very little CaM is unbound and available to bind to target proteins during changes in $[Ca^{2+}]_i$. Ruegg et al. (Ruegg et al. 1984) estimated that 10%, or roughly 3μM CaM was free in rehydrated smooth muscle fibers that had been lyophilized. Luby-Phelps et al. (1995) estimated that, at most, 5% was free, based on extraction from resting, permeabilized cultured tracheal smooth muscle cells, and on FRAP (fluorescence recovery after photobleaching) measurements in intact cells. Persechini and Cronk (1999) have used altered versions of the high-affinity MLCK CaM-binding domain as CaM probes in HEK-293 cells. They found a maximum unbound $(Ca)_4$-CaM concentration of approximately 45 nM at a free Ca^{2+} of 3 μM.

The issue of quantitating the level of available free CaM in the differentiated, contractile smooth muscle cell has relevance since at least two relatively low-affinity CaM-binding proteins in the smooth muscle cell have been postulated to play important regulatory roles. These two proteins, CaD and Ca-CaM-dependent protein kinase II (CaMKII), have K_ds for CaM at or above this estimate for available unbound CaM. The reported K_ds for CaD range from 50 to 430 nM, depending on the method of preparation (Zhuang et al. 1996). The reported K_d for CaMKII ranges between 15–200 nM at saturating [Ca], depending on the isoform studied (Braun and Schulman 1995; Kamm and Stull 1989; Schulman and Hanson 1993).

For low-affinity CaM binding targets such as CaMKII and CaD, the amount of available CaM in the cell is likely to be limiting. In contrast, for high-affinity CaM targets, such as MLCK (K_d~1 nM; Kamm and Stull 1985), CaM is not thought to be limiting and activity is regulated primarily by changes in $[Ca^{2+}]_i$. Recently, Hulvershorn et al.(2001) used a peptide taken directly from a low-affinity CaM binding domain of CaD and labeled with an environmentally-sensitive fluorophore as a probe of free [CaM]. The probe gave essentially no detectable fluorescence in vitro in the absence of CaM, but showed an increased fluorescence in the presence of CaM. On loading the peptide into intact differentiated vascular smooth muscle cells, significant signals were detected, indicating that sufficient free CaM was available to perturb this low-affinity CaM binding site. Additionally, significant changes in the fluorescence of the probe, and, presumably, the amount of available CaM in the cells were detected during activation of PKC by a phorbol ester. This indicates that the amount of available CaM in smooth muscle cells is a dynamic variable. Since α-adrenergic agonists have been reported to activate PKC in vascular smooth muscle (Merkel et al. 1991; Khalil et al. 1994; Aburto et al. 1995; Brozovich 1995; Ohanian et al. 1996; Suenaga and Kamata 2000; Sato et al. 2001), it is quite possible that there is an additional level of complexity for α-agonist-induced regulatory mechanisms, whereby α-agonists modulate the amount of CaM available for activation of low-affinity CaM targets.

Signal transduction leading to thick filament regulation

Myosin light chain kinase

General properties

It is clear that phosphorylation of the 20-kDa myosin light chain (LC20) by myosin light chain kinase (MLCK) represents a major regulatory pathway by which smooth muscle contraction is regulated. MLCK phosphorylates Ser19, and in the presence of extreme stimulation, Thr18 of the 20-kDa light chain of myosin. Phosphorylation at these sites results in increases in myosin ATPase activity and, hence, contraction.

MLCK represents a high-affinity target for CaM. The two high-affinity binding C-terminal Ca^{2+} binding sites of CaM are thought to contain bound Ca^{2+} at resting $[Ca^{2+}]_i$. In this state, CaM is thought to be bound to MLCK, but the kinase is not activated (Johnson et al. 1996; Bayley et al. 1996). An increase in $[Ca^{2+}]_i$ leads to the binding of Ca^{2+} to the low-affinity N-terminal binding sites of CaM and activation of MLCK via a conformational change. This results in the displacement of the autoinhibitory domain of MLCK from the catalytic site for LC20 binding and phosphorylation (Pearson et al. 1988).

Recently, it has become clear that N-terminal residues of MLCK also bind to actin filaments, whereas C-terminal residues bind to myosin. Approximate calculations of the size and shape of the actin/MLCK/myosin complex suggest that bound MLCK can indeed reach myosin while being bound to actin. The tethering of MLCK to actin may allow each MLCK molecule to phosphorylate multiple myosin molecules as crossbridge sliding ensues (Lin et al. 1999). This may compensate for the fact that the concentration of smooth muscle myosin LC20 is roughly 25 times greater than that of MLCK.

Ca²⁺ sensitivity of MLCK

MLCK was originally reported to be activated in a simple manner by $[Ca^{2+}]_i$ and CaM. Dephosphorylation of LC20 was initially thought to occur by the action of an unregulated phosphatase (Sobieszek 1977). However, when the luminescent $[Ca^{2+}]_i$ indicator aequorin was used in vascular smooth muscle, it was found that the $[Ca^{2+}]_i$ *sensitivity* of the contractile apparatus is a variable in intact muscles (Morgan and Morgan 1984). This finding has subsequently been confirmed with other Ca^{2+} indicators (Himpens et al. 1988), and in permeabilized preparations where $[Ca^{2+}]_i$ could be directly controlled (Nishimura et al. 1988). One mechanism by which the Ca^{2+} sensitivity of the contractile apparatus can be modulated, in theory, is by alteration of the Ca^{2+} requirement for activation of MLCK.

The Ca^{2+} sensitivity of MLCK can be regulated in vitro by phosphorylation of MLCK via a number of other kinases such as CaMKII, PAK, PKA, PKC, or ERK1/2 (Goeckeler et al. 2000; Horowitz et al. 1996a; Kim et al. 2000). With one exception, all of these phosphorylations are inhibitory, i.e., they decrease the Ca^{2+} sensitivity of MLCK. The exception is ERK1/2, which facilitates the activation of MLCK, i.e., increases the Ca^{2+} sensitivity of MLCK (Nguyen et al. 1999; Morrison et al. 1996).

Alpha-adrenergic agonists have been reported, in various vascular tissues, to activate both PKC and ERK1/2 (Xiao and Zhang 2002; Dessy et al. 1998). Thus far, however, there is no convincing evidence that an inhibitory phosphorylation of MLCK via PKC plays a physiologically significant role in the regulation of MLCK activity in vascular smooth muscle. Nevertheless, depolarization-induced activation of ERK has been associated with an increased LC20 phosphorylation and contraction in vivo in ferret aortic strips (Kim et al. 2000). Whether this pathway is recruited in the presence of α-agonists is not known. Alpha agonist-induced contractions of ferret aortic strips are associated with very low steady state values of LC20 phosphorylation, which makes it unlikely that ERK1/2 plays a major role in activating MLCK under these conditions. In other tissues (see next section), sustained elevations of LC20 phosphorylation have been reported in the presence of α-agonists and regulation by this pathway is a possibility.

Ca-independent myosin light chain kinase

A recent series of studies from the Walsh lab has identified a Ca^{2+}-independent MLCK in rat tail artery. Given that the tail artery is often used as a model of resistance arteries, the pathways described may be relevant. In this vessel, a relatively high resting level of LC20 phosphorylation and a further increase in LC20 phosphorylation induced by inhibitors of MP were both found to be resistant to the inhibitors of MLCK, ML-9 (Mita and Walsh 1997) and AV25 (Weber et al. 1999). The Ca^{2+}-independent MLCK activity was identified as a smooth muscle form of integrin-linked kinase (ILK) that coassociates with myosin filaments (Deng et al. 2001). The authors suggested that ILK might contribute to increases in LC20 phosphorylation that occur in response to activation of receptors coupled to Gq/11 or G12/13 heterotrimeric GTP-binding proteins; however, direct evidence supporting this suggestion is still lacking.

Additionally, in the above study (Deng et al. 2001), an α-agonist (cirazoline)-induced an increase in LC20 phosphorylation levels above baseline, but this increase was sensitive to antagonists of MLCK. Thus, it seems unlikely that α-adrenergic agonists would utilize this newly discovered ILK pathway.

Myosin phosphatase

General properties

Myosin phosphatase (MP) is composed of three subunits: a catalytic subunit (PP1c), a large noncatalytic subunit (targeting subunit, MYPT1, M130), and a small noncatalytic subunit of unknown function (M20; Hartshorne et al. 1998). The binding of MYPT1 to PP1c increases the activity of the catalytic subunit towards phosphorylated myosin (Ichikawa et al. 1996b). Thus, MYPT1 plays an activating role, but it has also been suggested to play a targeting role (Alessi et al. 1992) in the regulation of MP. Although binding of MP to myosin has been demonstrated in fractionated cells (Johnson et al. 1997) and in vitro (Ichikawa et al. 1996b), imaging studies of intact cells have raised some questions on this issue. In differentiated, contractile ferret portal vein smooth muscle cells, the association of the targeting subunit with the catalytic subunit of MP is agonist-specific. During steady state PGF2α-induced contractions, MYPT1 was located at the cell surface during steady state contraction, whereas PP1c was found in the core of the cell. The lack of colocalization of MYPT1 and PP1c was associated with a sustained, high level of LC20 phosphorylation, even though PGF2α has previously been shown to cause little or no increase in $[Ca^{2+}]_i$ (Suematsu et al. 1991), implying that the dissociation of the subunits caused an inhibition of phosphatase activity. In the presence of the α-agonist, PE, however, no such dissociation of the subunits was seen, consistent with the low steady state levels of LC20 phosphorylation observed and the expected high activity of MP (Shin et al. 2002).

MP regulation during α-adrenergic-induced contractions

Considerable effort is currently being expended to investigate mechanisms by which inhibition of MP might occur. In theory, an agonist-induced inhibition of MP would lead to sustained increases in LC20 phosphorylation and, consequently, sustained elevations of contractile force. It should be kept in mind, however, that as originally described by Dillon et al. (1981), sustained elevations of force in smooth muscle are often associated with falling LC20 phosphorylation levels. This phenomenon was originally described as "latch" by the Murphy group (Hai and Murphy 1989).

Alpha agonists have specifically been shown to cause sustained contractions, accompanied by falling LC20 phosphorylation levels in large vessels (Jiang and Morgan 1989) and in rat cremaster muscle arterioles (Zou et al. 2000). Similarly, NE causes transiently elevated LC20 phosphorylation levels in canine anterior tibial arteries in vivo during maintained elevation of vascular resistance (Moreland et al. 1990). It is of note that in the rat tail artery, a vessel often used as a model of resistance arteries, sustained elevations of LC20 phosphorylation have been reported in response to the α-agonist cirazoline (Mita and Walsh 1997). Thus, the relative importance of regulation of MP activity during α-agonist-induced contraction of resistance vessels is currently unclear and may well be vessel-specific.

Rho kinase

Rho kinase (ROK, or ROCK) is a serine/threonine kinase activated by the GTP-bound form of *rho*-A. Activated ROK phosphorylates MYPT1 at T695 and inhibits MP activity (Kimura et al. 1996). There is a strong case in the literature that ROK is the physiologically important MP kinase and the primary mode of regulation of MP, based on extensive biochemical and pharmacological experiments (Ichikawa et al. 1996a; Kimura et al. 1996; Feng et al. 1999; Nobe and Paul 2001; Nagumo et al. 2000; Bolz et al. 2000; Somlyo and Somlyo 1998; Fig. 4). Recently, however, this model has been called into question by the suggestion of MacDonald et al. (2001) that the endogenous kinase of Ichikawa et al. (1996a) is actually a ZIP-like kinase. These authors have suggested that the ZIP-like kinase directly phosphorylates MYPT1 and that it functions downstream of ROK (MacDonald et al. 2001). This issue remains controversial.

A large fraction of the evidence supporting a major role, either directly or indirectly, for ROK has been based on inhibition of contraction by the ROK inhibitor Y 27632, or in fewer cases, HA-1077. Recent studies have made it clear that Y27632, although very selective in vitro (Uehata et al. 1997), is not entirely selective in vivo (K.G. Morgan, unpublished observations). Thus, caution needs to be applied in interpreting these studies, especially in the many cases where LC20 phosphorylation is not directly measured to confirm the putative effect of the ROK inhibitor to antagonize MP.

Separate from the possibility of directly phosphorylating MYPT1, ROK has been implicated in several additional pathways. ROK has been suggested to be upstream of CPI-17-mediated inhibition of MP (Kitazawa et al. 2000), to directly phosphorylate LC20 (Amano et al. 1996; Kureishi et al. 1997), to phosphorylate the actin binding protein CaP (Kaneko et al. 2000) and to be required for activation of ERK1/2 (Matrougui et al. 2001). Specifically with regard to resistance vessels, Matrougui et al. have presented evidence that in rat mesenteric resistance vessels, ROK was required for angiotensin-induced activation of ERK1/2 . ERK1/2 activation has been associated with α-adrenergic agonist-induced contractions of blood vessels, but whether this is linked to ROK activation is not known.

A translocation of ROK and *rho*-A to the cell membrane has been shown to occur with agonist activation (Taggart et al. 1999; Somlyo and Somlyo 2000). How ROK, located at the plasmalemma, might regulate the dephosphorylation of myosin in the core of the cell is problematic, although it may be involved in the membrane targeting MP seen in ferret vascular smooth muscle (Shin et al. 2002). However, the membrane association of ROK also raises the question of whether ROK might have functions near the cell membrane distinct from any direct or indirect effect on myosin filaments in the core of the cell. Recent-

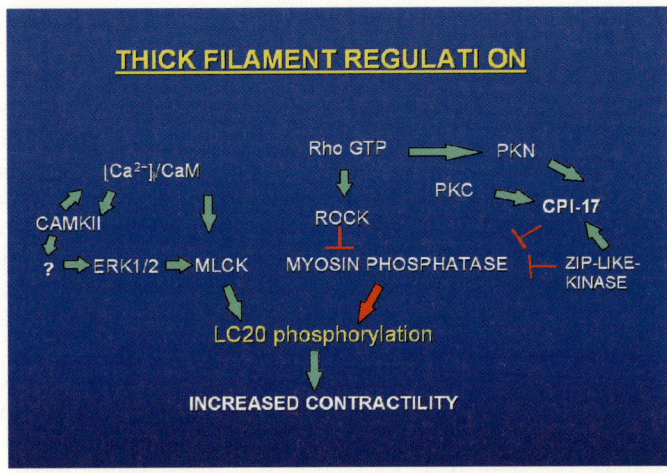

Fig. 4 Mechanism of thick-filament regulation. A simplified scheme illustrating the major pathways of thick filament regulation for which there is appreciable experimental evidence in vascular smooth muscle. See "Abbreviations." *Green* represents activation, *red* represents an inhibitory influence. In general, for crossbridges to cycle and contraction to occur, the 20-kDa myosin light chain (LC20) must be phosphorylated. Phosphorylation of LC20 is regulated by a kinase-phosphatase couple, myosin light chain kinase (*MLCK*) and myosin phosphatase (*MP*). Thus, through regulation of MLCK and MP by pathways other than activation of MLCK by Ca^{2+}/Calmodulin alone, the apparent Ca^{2+} sensitivity of contraction may be modulated. For example, pathways that lead ultimately to inhibition of MP activity increase the apparent Ca^{2+} sensitivity of contraction. MLCK is activated primarily by Ca^{2+}/Calmodulin, and possibly also by ERK1/2, although the role of ERK1/2 in α_1-agonist induced contractions remains somewhat uncertain (see the section entitled "Extracellulary regulated kinase (ERK1/2)" for discussion). The activity of MP is inhibited by a pathway involving *rho* kinase (*ROCK*). ZIP-like kinase directly phosphorylates MYPT1 (inhibitory) and it may function downstream of ROCK. CPI-17 is a smooth muscle-specific protein and an inhibitor of MP. Phosphorylation of CPI-17 at Thr 38 by PKC activates CPI-17 to become an inhibitor of the catalytic activity of MP

ly, it was shown by imaging studies that agonist-induced myosin phosphorylation in primary cultured tracheal smooth muscle cells was inhibited by the ROK inhibitor Y27632 only in the submembranous cortex and not in the core of the cell, where the bulk of the contractile/stress fibers are located (Miyazaki et al. 2002). In fact, a considerable body of work suggests that ROK may play an important role in the regulation of mechanical plasticity and synthetic activity of the smooth muscle cell (Halayko and Solway 2001). ROK has also been implicated in the etiology of a number of proliferative vascular disorders. Notably, a dominant negative version of ROK, introduced into a pig model of arteriosclerosis by adenovirus-mediated transfer, led to a reversal of vessel narrowing and vasospastic contraction characteristic of this model (Morishige et al. 2001).

Regarding the specific question of the degree to which ROK might be involved in α-adrenergic agonist-induced contraction of microvessels, little is known. Some effect of Y27632 on PE-induced contractions have been observed in large vessels, but, consistently, the inhibition of α-agonist contractions has been noted to be far smaller than that of the thromboxane mimetic U-46619 (Sakurada et al. 2001) and PGF2α (Shin et al. 2002).

CPI-17

CPI-17 is a smooth muscle-specific protein (Eto et al. 1997) and an inhibitor of MP (Eto et al. 1995). Phosphorylation of CPI-17 at Thr 38 by PKC activates CPI-17 to become an inhibitor of the catalytic activity of MP. The effect appears to be a direct inhibition of the catalytic subunit of MP, PP1c. (Eto et al. 1997; Senba et al. 1999). It has also been suggested that the fatty acid- and Rho-activated kinase, PKN, might also be able to increase the inhibitory potential of CPI-17 through a similar phosphorylation (Hamaguchi et al. 2000). Selective depletion of CPI-17 by triton permeabilization of smooth muscle diminishes PKC-induced Ca^{2+}-sensitization of femoral artery strips (Kitazawa et al. 1999). Woodsome et al. (2001) have shown that the ratio of expression CPI-17 to that of MP correlates with the ability of PKC activators to cause $[Ca^{2+}]_i$ sensitization.

Since there is considerable evidence that α-adrenergic agonists activate PKC, that PKC inhibitors can inhibit smooth muscle contraction, and that phorbol esters activate contraction (e.g., Chatterjee et al. 1986; Danthuluri and Deth 1984; Jiang and Morgan 1987; Khalil and Morgan 1993; Lee et al. 1999; Rasmussen et al. 1987; Singer 1990), it is feasible that α-agonist-induced activation can lead to activation of CPI-17, inhibition of MP, an increase in LC20 phosphorylation, and contraction of the tissue (Fig. 4). If this pathway plays a significant role in α-adrenergic agonist-induced contractions, the prediction would be that MP inhibition would produce sustained contractions accompanied by sustained elevations of LC20 phosphorylation levels, although it is not yet clear whether or not this actually occurs.

CaMKII

General properties

CaMKII is a multifunctional serine/threonine protein kinase originally isolated from brain (reviewed in, Braun and Schulman 1995; Singer et al. 1996; Soderling et al. 2001) and now known to be encoded by four genes (α, β, γ, δ). Each gene product can be alternatively spliced to form multiple additional isoforms. The α and β isoforms are primarily expressed in neural tissue, but the γ and δ isoforms are expressed in most tissues, including vascular smooth muscle (Singer et al. 1997; Kim et al. 2000). Each monomer consists of an N-terminal catalytic region and a central regulatory domain, including an autoinhibitory domain and a Ca^{2+}/CaM binding domain, followed by several variable regions and a C-terminal association domain. Each of the monomers has a molecular weight of 50–65 kDa and, through the association domains, forms large multimers of 8–12 subunits, which, for the α subunit, are organized as two stacked rings in a "hub-and-spoke" (or, "gear and foot") arrangement where the association domain is in the center and the catalytic domains are arranged in the periphery (Soderling et al. 2001).

CaMKII undergoes an autophosphorylation reaction that has two consequences: (a) The off rate for the dissociation of CaM slows by several orders of magnitude and hence the affinity for CaM appears to greatly increase, and (b) the catalytic domain develops an autonomous Ca/CaM-independent kinase activity. Thus, the molecule develops a "memory" of having previously seen Ca/CaM (Soderling et al. 2001). The effect of $[Ca^{2+}]_i$ signals on CaMKII has been shown to be frequency and duration dependent; thus it has been suggested that CaMKII serves as a decoder of repetitive $[Ca^{2+}]_i$ signals (De Koninck and Schulman 1998).

Isoform-specific properties

Isoform-specific in vitro properties have been reported (Naito et al. 1997) and provide a possible mechanism for producing isoform-specific functions and targets in the smooth muscle cell (Singer et al. 1997; Srinivasan et al. 1994; Brocke et al. 1995; Braun and Schulman 1995). The affinity of CaMKII fo Ca/CaM varies between isoforms. For example, the β isoform has a greater affinity than the α isoform. The half maximal [CaM] for activation of CaMKII has been reported to range widely, from 15–200 nM in vitro (Schulman and Hanson 1993; Brocke et al. 1999; Colbran and Soderling 1990). The β isoform contains an actin binding domain. Alternatively spliced variants of the brain δ isoform (Takeuchi et al. 1999) contain a nuclear localization signal. Thus far, similar isoform-specific domains or functions have not been described for the γ or δ isoforms present in contractile smooth muscle.

Role of CaMKII in smooth muscle

Whether CaMKII is recruited during α-adrenergic agonist-induced contractions is not yet clear. However, it seems likely that any agonist that induces $[Ca^{2+}]_i$ waves of sufficient amplitude, duration, and frequency should activate CaMKII. Furthermore, since α-agonists are known to activate PKC, which may upregulate the available [CaM], it seems feasible that α-agonist activation may lead to CaMKII activation in some settings.

Recently, an antisense approach was used to knock down CaMKII protein levels in vascular smooth muscle (Kim et al. 2000). A nonisoform-specific sequence was targeted against a highly conserved catalytic domain sequence. The functional effect was a significant decrease in KCl-induced contractions, and a decrease in KCl-induced LC20 phosphorylation. This indicates that endogenous CaMKII enhances smooth muscle contractility through thick filament regulation, at least during the increase in $[Ca^{2+}]_i$ that occurs during KCl contractions. In the same study, activation of PKC with a phorbol ester produced a contraction that was insensitive to the decrease in CaMKII protein levels. Similar effects were seen with a pharmacological approach (Rokolya and Singer 2000; Kim et al. 2000).

CaMKII is also reportedly involved in signaling cascades leading to an activation of ERK1/2 in both cultured smooth muscle cells and contractile smooth muscle tissue (Abraham et al. 1997; Kim et al. 2000). ERK1/2 phosphorylates MLCK in a manner that activates the kinase (Nguyen et al. 1999). Thus, a pathway may exist in smooth muscle linking increases in $[Ca^{2+}]_i$ to CaMKII activation, subsequent ERK1/2 activation, MLCK activation, and LC20 phosphorylation (Fig. 4).

In contrast with the above studies, CaMKinase II has previously been suggested to inhibit contractility in tracheal smooth muscle by directly phosphorylating MLCK at an inhibitory site. This results in a negative regulation of LC20 phosphorylation and contractile force (Kamm and Stull 1985). Thus, there may be tissue-specific differences in the actions of CaMKII.

CaMKII has also been reported to be involved in the activation of Ca^{2+} channels (McCaron et al. 1992), and inhibition of Ca^{2+}-activated Cl channels(Greenwood et al. 2001).

Additionally, SR Ca^{2+}-ATPase activity in smooth muscle has been reported to be activated by CaMKII (Grover et al. 1996). More recently, it has been suggested that the functional effect of CaMKII to accelerate the rate of decline of $[Ca^{2+}]_i$ in smooth muscle is due to an effect on mitochondria rather than the SR (McGeown et al. 1998). Given that pharmacological antagonists and antisense oligos directed against CaMKII are reported to be inhibitory with respect to tonic KCl contractions in smooth muscle, it is possible that the effects on Ca^{2+} channels are involved, at least in part, in these effects. The effects on the rate of decline of $[Ca^{2+}]_i$ transients may be more relevant to phasic contractions.

CaMKII may also be involved in thin filament regulation (next section), as it has been shown to phosphorylate CaP, as well as CaD in vitro. The phosphorylation of CaP reverses its inhibition of myosin ATPase activity in vitro (reviewed in, Horowitz 1996b). The N-terminal phosphorylation of CaD by CaMKII prevents the "tethering" of actin to myosin by CaD (Sutherland and Walsh 1989). Whether these in vitro interactions have physiological relevance is currently unknown. A significant fraction of CaMKII associates with a myofibrillar fraction in smooth muscle (Singer et al. 1996) and copurifies with the smooth muscle thin filament protein, CaD (Ngai and Walsh 1984). These biochemical studies suggest that in situ, the kinase might codistribute with the cytoskeletal domain, but, conversely, the functional effects on Ca^{2+} channels and SR Ca^{2+}-ATPase activity mentioned above suggest a distribution of the kinase with the membrane or the SR. Because of the lack of adequately selective antibodies, direct imaging studies have not yet been performed in contractile smooth muscle, but such studies may well reveal agonist-induced translocations of CaMKII to explain these apparently disparate results.

Signal transduction leading to thin filament regulation

It is clear that smooth muscle crossbridges are regulated by phosphorylation of the regulatory light chains of myosin (LC20). It is also quite clear that changes in contractile force and crossbridge cycling rates often occur in the absence of corresponding changes in LC20 phosphorylation levels and that tone is often maintained in the presence of falling LC20 phosphorylation levels (e.g., Siegman et al. 1984; Gunst et al. 1992, 1994; Dillon et al. 1981). Furthermore, in the early 1980s it was demonstrated that there is a calcium sensitivity of myosin ATPase activity in preparations containing skeletal muscle myosin and smooth muscle thin filaments (Marston and Smith 1984; Marston et al. 1980), suggesting the existence of Ca-sensitive smooth muscle actin binding proteins. Additionally, there is evidence that under some conditions, unphosphorylated smooth muscle crossbridges are not completely turned off (Haeberle 1999; Sellers 1999; Wagner and George 1986), providing a functional need for an actin regulatory system, in addition to thick filament regulation in smooth muscle. Although there is no troponin in smooth muscle, two proteins have been suggested to possibly fulfill an analogous function, CaD and CaP (Fig. 5).

Caldesmon

Caldesmon (CaD) is an actin, tropomyosin, myosin and CaM binding protein. The CaD family contains two subfamilies: smooth muscle CaDs of a higher molecular weight (793 residues long), and CaDs with a lower molecular weight that are found in both muscle and nonmuscle cells. CaD's putative role in smooth muscle function has recently been reviewed (Morgan and Gangopadhyay 2001). C-terminal domains of CaD are responsible for most of the demonstrated in vitro functions of CaD, including actin binding and inhibition of myosin ATPase activity. The N-terminal portion of the molecule binds myosin in vitro (Chalovich 1988; Hemric et al. 1994; Wang et al. 1997; Li et al. 2000; Ikebe and Reardon 1988). This raises the possibility that by simultaneously binding actin and myosin, CaD might tether myosin and actin together. Peptides against the myosin-binding domain of CaD selectively inhibit α-agonist-induced contractions (Lee et al. 2000), consistent with a role of CaD in α-agonist-induced signal transduction. It has been proposed that the tethering of CaD to both actin and myosin has the effect of positioning CaD in the contractile filaments and stabilizing the position of actin relative to myosin.

Two main approaches have been used to probe the function of endogenous CaD in vivo. Firstly, a peptide antagonist of CaD (GS17C) has been made, targeted to Gly651-Ser667 of the gizzard CaD sequence. This part of the CaD sequence contains a high-affinity actin binding site, thought to be important in properly positioning downstream C-terminal actin binding sites of CaD over myosin's actin binding site. In itself, however, GS17C has no direct myosin inhibitory actions; thus, it functions in the manner of a dominant negative construct. When introduced into contractile vascular smooth muscle cells, GS17C induces a sustained elevation of basal tone (Katsuyama et al. 1992). More recently, identical results were obtained for esophageal smooth muscle (Sohn et al. 2001). These results are consistent with a role of endogenous CaD to suppress basal tone.

A second approach used to test the endogenous function of CaD in smooth muscle has been to use a chemical loading procedure (Morgan and Morgan 1982) to introduce antisense against CaD into smooth muscle strips (Earley et al. 1998). Downregulation of CaD protein levels caused a sustained elevation of basal contractile tone, consistent with the peptide studies and consistent with a role of endogenous CaD to tonically suppress contractile tone.

CaD's endogenous action to inhibit myosin ATPase activity, and as a result, contractile tone, is thought to be regulated by the binding of Ca^{2+}-CaM and/or the phosphorylation of sites between the two C-terminal actin binding domains (Morgan and Gangopadhyay 2001).

Calmodulin

The binding of Calmodulin (CaM) to CaD is known to reverse the inhibition of myosin ATPase activity caused by CaD in vitro. The CaM binding sites of CaD have a relatively low affinity for CaM and it has been questioned whether sufficient free CaM is available to cause a significant change in CaD's function. However, the amount of CaM available for binding to CaD may be a dynamic variable and more recent measurements of the affinity of CaM for CaD has resulted in higher estimates (Hulvershorn et al. 2001). It has also been reported that a modified version of CaM exists in smooth muscle that has a significantly higher affinity for CaD than does unmodified CaM (Notarianni et al. 2000). As discussed above, the possibility exists that α-adrenergic agonists regulate both $[Ca^{2+}]_i$ and the levels of available CaM in smooth muscle cells. Thus, α-adrenergic agonist-induced contractions may well involve a disinhibition of CaD's effects on actomyosin interactions via a Ca^{2+}/CaM pathway (Fig. 5).

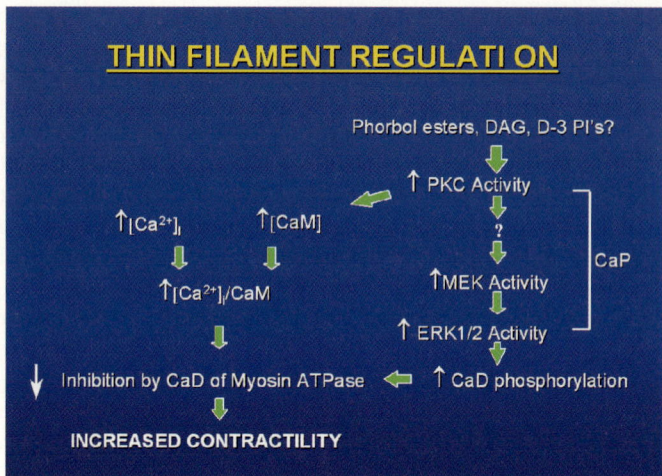

Fig. 5 Mechanisms of thin-filament regulation. Simplified scheme illustrating pathways that have been linked with α_1-adrenoceptor-induced activation through thin filament mechanisms. See "Abbreviations." While smooth muscle crossbridges are regulated primarily by phosphorylation of the regulatory light chains of myosin (*LC20*), changes in contractile force and crossbridge cycling rates can occur in the absence of corresponding changes in LC20 phosphorylation. Under some conditions, unphosphorylated smooth muscle crossbridges are not completely turned off. (see the section entitled "Signal transduction leading to thin filament regulation" for discussion) This may be viewed as providing a functional need for an actin regulatory system, in addition to thick filament regulation in smooth muscle. Although there is no troponin in smooth muscle, two proteins have been suggested to possibly fulfill an analogous function, caldesmon (*CaD*) and calponin (*CaP*). It has been proposed that the tethering of CaD to both actin and myosin has the effect of positioning CaD in the contractile filaments and stabilizing the position of actin relative to myosin. Phosphorylation of CaD may have the effect of "disinhibiting" myosin ATPase activity. α_1-adrenergic agonist-induced contractions may involve a disinhibition of CaD's effects on actomyosin interactions via a Ca^{2+}/CaM pathway. In some smooth muscles, α_1-adrenergic agonist-induced activation of ERK1/2 may lead to CaD phosphorylation, and disinhibition of myosin ATPase activity. The role of calponin (*CaP*) is controversial. It has been proposed by different groups that CaP may function either in a troponin-like manner to directly inhibit the actin-activated Mg-ATPase activity of myosin or as a signaling molecule to facilitate PKC-dependent ERK1/2 activation. The data supporting the two hypotheses have been recently reviewed. (Morgan and Gangopadhyay 2001)

Extracellularly regulated kinase

Considerable evidence exists that α-adrenergic agonists cause activation of extracellularly regulated kinase (Erk1/2) in vascular smooth muscle (Xiao and Zhang 2002; Je et al. 2001; Khalil et al. 1995; Xu et al. 1996; Roberts 2001) and CaD has been suggested to be a likely target of activated ERK1/2. It is also possible that the effects of CaM on CaD may be synergistic with those of ERK1/2-mediated phosphorylation of CaD.

The potential role of ERK1/2 in thin filament regulation has been recently reviewed (Morgan and Gangopadhyay 2001). In mammalian smooth muscle, there are two phosphorylation sites for ERK1/2 in the C-terminal end of CaD, S759 (homologous to gizzard S702) and S789. Using an antibody that detects a signal when either of the ERK1/2 sites on CaD is phosphorylated, a time- and agonist-dependent three- to fivefold increase in signal during α-agonist-induced contraction of ferret aorta has been detected (Dessy et al. 1998). Yamboliev et al., using site-specific antibodies, have seen phosphorylation at S789 but not at S759 in vascular, airway, and gastrointestinal smooth muscles (Yamboliev et al.

2000; Hedges et al. 2000; Cook et al. 2000). ERK1/2 has also been reported to colocalize with CaD during α-agonist-induced contractions (Khalil et al. 1995). Thus, taken together, the available evidence suggests that in some tissues α-adrenergic agonist-induced activation of ERK1/2 may lead to CaD phosphorylation (Fig. 5).

Furthermore, although the topic is controversial, some studies suggest that CaD phosphorylation at ERK1/2 sites leads to a reversal of the inhibitory actions of CaD on actomyosin interactions. When CaD is phosphorylated by ERK1/2, CaD no longer inhibits movement in a motility assay (Adam et al. 1997; Gerthoffer et al. 1996). ERK1/2 phosphorylation of CaD also prevents actin binding (Li et al. 2001; Marston et al. 2001). The MEK inhibitor PD098059 has been used to prevent the activation of ERK1/2, and mixed results have been reported. Some groups have reported negative results (e.g., Gorenne et al. 1998), but others (e.g., Watts 1996; Dessy et al. 1998) have reported positive results. Thus, it is possible that ERK1/2 may well be a physiological CaD kinase, at least in some types of vascular smooth muscle, but tissue-specific effects may complicate the situation.

On the other hand, the possibility remains that ERK1/2 may have other targets in the smooth muscle cell. KCl (see "Extracellularly regulated kinase ERK1/2") or endothelin-mediated activation can lead to an ERK1/2-mediated activation of MLCK (Kim et al. 2000; D'Angelo and Adam 2002). It is also well known that ERK1/2 can regulate immediate early gene expression through an effect on transcription factors in the nucleus (Abe et al. 2000; Whitmarsh and Davis 1996); however, although ERK1/2 has been seen to translocate to the intranuclear space in primary cultured smooth muscle cells activated with an α-agonist (Mii et al. 1996), a nuclear translocation of ERK1/2 in differentiated contractile smooth muscle cells does not appear to occur (Khalil and Morgan 1993). Thus, ERK1/2 may play a role in more than one signal transduction pathway in contractile and synthetic smooth muscle.

Calponin

General properties

Calponin (CaP) is a relatively recently discovered 32–36-kDa smooth muscle-specific protein, whose function is still controversial. Basic h1 CaP was originally identified as a marker of differentiated smooth muscle (Takahashi et al. 1988; Gimona et al. 1990). All CaP isoforms appear to interact with F-actin and inhibit actomyosin Mg-ATPase activity in vitro (for review, see Gimona and Small 1996). CaP binds to CaM (Wills et al. 1993), myosin (Shirinsky et al. 1992), desmin (Mabuchi et al. 1997), and phospholipids (Bogatcheva and Gusev 1995) but the physiological significance of these interactions is not yet clear. CaP is a substrate for PKC and Ca/CaM-dependent kinase II in vitro (Winder and Walsh 1990).

Basic CaP contains a single calponin homology (CH) domain at the N-terminal end, followed by an actin binding region (Mezgueldi et al. 1995) referred to as the ABS-1 domain (for actin binding sequence no. 1). This is followed by three C-terminal repeats. The C-terminal repeats are also reported to bind actin (Mino et al. 1998). The function of CH domains in proteins is somewhat unclear as the domain is present in actin binding proteins as well as in signaling proteins (Gimona and Mital 1998). In single CH domain containing proteins, such as calponin, SM22, IQGAP and Vav, the CH domain sequence does not appear to bind actin (Gimona and Winder 1998). The CH domain of CaP is an ERK binding domain (Leinweber et al. 1999). Additionally, a central segment of CaP containing part of the TNI-like domain and the first of the C-terminal repeats binds PKCs regulatory domain and facilitates the activation of PKC (Leinweber et al. 2000).

Function of CaP in smooth muscle

The role of calponin in muscle contractility is controversial. It has been proposed by different groups that CaP may function either in a troponin-like manner to directly inhibit the actin-activated Mg-ATPase activity of myosin (Winder and Walsh 1990) or as a signaling molecule to facilitate PKC-dependent ERK1/2 activation (Menice et al. 1997). The data supporting the two hypotheses have been recently reviewed (Morgan and Gangopadhyay 2001). It seems quite likely that the protein can perform different functions in different smooth muscle tissues under different conditions.

Regarding α-agonists, however, in the early time points after agonist activation in ferret vascular smooth muscle, ERK and PKC coimmunoprecipitate and cotranslocate with CaP to the vicinity of the plasmalemma (Menice et al. 1997). This suggests a role for CaP as a signaling molecule. CaP facilitates activation of PKC and links targeting of ERK and PKC to the surface membrane, in the manner of an adaptor protein (Leinweber et al. 1999; Menice et al. 1997).

A CaP knock-out mouse has been reported (Yoshikawa et al. 1998). The most obvious phenotypic change observed was, surprisingly, an increase in bone formation. The lack of obvious smooth muscle defects may be related to compensatory changes in other contractile proteins. Smooth muscles from the calponin knock-out mice had actin levels that were decreased by 25–50%. This dramatic change in actin levels alone may explain the slightly elevated unloaded shortening velocity observed in some phasic smooth muscle tissues of these animals (Matthew et al. 2000). Alterations in desmin and CaD were also seen in these mice.

On the other hand, an antisense approach has also been used to produce blood vessels in which CaP levels are acutely decreased by 50% with no change in other cytoskeletal protein levels (Je et al. 2001). The effect of the knock-down of CaP levels in these vessels was a specific inhibition of α-agonist-induced contractions accompanied by an inhibition of agonist-induced phosphorylation of ERK1/2 and CaD. Thus, these results are consistent with a role of CaP in facilitating α-agonist-induced signal transduction (Fig. 5).

Other mechanisms of signal transduction

Cytoskeletal remodeling

It is known that actin filament rearrangements are regulated in growing smooth muscle cells. It has also been suggested that actin filament remodeling may play a role in contractile smooth muscle (Gunst and Tang 2000). If actomyosin interactions initially cause the development of force and a strain on the cytoskeleton, a "gluing" of actin filaments together and to sites where they insert into dense bodies and attachment plaques, then this could maintain tension in the absence of further crossbridge cycling. This mechanism is appealing since it could explain the well-known property of smooth muscle to be able to maintain tone in the absence of ATP utilization (i.e., "latch"). Additionally, disruption of actin filaments has been associated with an inhibition of agonist-induced $[Ca^{2+}]_i$ signals (Tseng et al. 1997), providing a possible mechanism by which mechanical signals could alter Ca signaling.

Latch

The latch hypothesis of Hai and Murphy (1989) was developed to explain the observed dissociations between LC20 phosphorylation and tension levels during tonic contractions. This hypothesis assumes that the dephosphorylation of an attached crossbridge results in a reduction in the detachment rate of that crossbridge, i.e., the formation of a "latch-bridge." The resulting four-state kinetic model (Hai and Murphy 1989) can be fit to many experimental data with the appropriate choices for rate constants. However, clear-cut exceptions do exist (e.g., Nishimura et al. 1988; Jiang and Morgan 1987, 1989; Tansey et al. 1990; Gerthoffer et al. 1989; Moreland et al. 1987, 1992; Bradley and Morgan 1987; Papageorgiou and Morgan 1991). It has further been suggested that dephosphorylation of attached crossbridges in a strongly bound AM-ADP state, combined with "cooperative" reattachment of nonphosphorylated crossbridges through an action of attached crossbridges, could contribute to a modified latch model (Somlyo et al. 1988; Vyas et al. 1992). However, in these proposals it is difficult to visualize how the muscle maintains the stable steady-state levels of tone that are typical of tonic vascular smooth muscle, and it seems likely that latch would lead to irreversible, continuously increasing contractions (Murphy 1994). Furthermore, since the biochemical correlate of the latch-bridge has not been clearly identified, it is also unknown if adrenergic mechanisms regulate the formation of latch-bridges.

Integration

In the preceding sections we presented the phenomenology of Ca^{2+} signaling during α_1-adrenoceptor activation and the phenomenology of the biochemical systems that may also be involved. Experiments designed to elucidate the characteristics of protein kinase signaling cascades must, of necessity, be performed in tissue that is not functioning in the usual physiological manner. For example, production of force at a constant level of intracellular $[Ca^{2+}]$ in an α-toxin permeabilized muscle exposed to GTPγS shows unequivocally the existence and characteristics of G-protein-linked systems that influence Ca^{2+} sensitivity, but do not reveal how such systems actually operate in a physiological, agonist-induced contraction, in which Ca^{2+} signaling is localized, changing over time, and in which other, interacting signal transduction systems are also activated. In this section, therefore, we attempt to integrate the available data and review certain recent experiments, conducted under near physiological conditions, which do allow a certain level of integration of information. The biochemistry of thick and thin filament regulation is summarized in Figs. 4, 5.

We have reviewed primarily the available data on adrenergic activation of mammalian arterial smooth muscle. Yet, even within this category of smooth muscle, significant differences exist. A limitation on our ability to present a comprehensive picture of the contractile mechanisms activated by α_1-adrenoceptor stimulation in arterial muscle is that biochemical studies have been performed mainly on the smooth muscle of large arteries, such as ferret aorta or swine carotid artery, (as the biochemistry is facilitated by the availability of large amounts of tissue), but the high resolution imaging of Ca^{2+} has been performed mainly on the small arteries (as they present a more desirable specimen, optically). Furthermore, the physiology of these two types of arteries is likely to be significantly different from each other, reflecting their very different functions and means of physiological regulation. Therefore, in our summary, we distinguish these two types of arterial smooth muscle. Our intent is to advance a scenario of the most likely and important events taking place within these two types of arteries, after α_1-adrenergic activation.

Large arteries

In large arteries, spatially averaged $[Ca^{2+}]_i$ falls dramatically from the peak transient levels and is low during force maintenance (e.g., Menice et al. 1997). This fact, and the fact that MLC_{20}-P levels also fall dramatically during the maintenance of force, suggest a minor importance for mechanisms that increase Ca^{2+} sensitivity by inhibiting MLCP or stimulating MLCK; operation of these would enhance MLC_{20}-P. Thus, there is little evidence that CPI-17 or the ROK are important in α_1-adrenoceptor-mediated contractions in this tissue. Similarly, other mechanisms such as CaMKinase II-dependent MAPK phosphorylation of MLCK should be relatively unimportant as this would also increase $[MLC_{20}$-P]. These facts lead us to the conclusion that force must be maintained in large arteries mainly by thin-filament regulation. The available data support the idea, proposed previously (Morgan and Gangopadhyay 2001), that occupation of α_1-adrenoceptors leads to activation of εPKC, which binds CaP, which binds ERK. The three proteins translocate to the cell membrane, where ERK codistributes with MEK and is phosphorylated. Phosphorylation of ERK releases ERK from the membrane and targets ERK to CaD on the actin filaments. Phosphorylation of CaD by ERK contributes to disinhibition of crossbridge cycling (Morgan and Gangopadhyay 2001; Je et al. 2001) although the possible contribution of the latch-bridge phenomenon to the contraction of these arteries is still unknown.

Small arteries—rat mesenteric artery

Small arteries, such as rat mesenteric small arteries or rat cerebral arteries, have recently been studied under near physiological conditions, including intraluminal pressure and mammalian temperature. The use of pressurized (isobaric) arteries, as opposed to isometric arteries (i.e., mounted as rings on wires) is essential. Sensitivity to agonists, which may be a physiological manifestation of increased Ca^{2+} sensitivity, is higher in isobaric than in isometric small arteries (Buus et al. 1994; Dunn et al. 1994). Similarly, Ca^{2+} sensitivity of pressurized rat cerebral arteries is enhanced at 37°C, as opposed to 22°C (Gokina and Osol 1998). Most importantly, under these conditions, small arteries develop significant myogenic tone, which does increase Ca^{2+} sensitivity. For example, the Ca^{2+} sensitivity of pressurized rat mesenteric small arteries was found to be five times higher during pressure-induced activation compared with potassium stimulation, and twice as high as the sensitivity during α_1-adrenergic activation (Van Bavel et al. 1998). The Rho-kinase pathway appears to increase Ca^{2+} sensitivity significantly in pressurized rat mesenteric small arteries (Van Bavel et al. 1998) and activation of PKC appears to constrict pressurized cerebral arteries by an increase in Ca^{2+} sensitivity (Gokina et al. 1999). Of course, abundant evidence links PKC to agonist-induced vasospasm of small arteries (Batchelor et al. 2001) and ROK to pathological states such as hypertension (Mukai et al. 2001). For present purposes however, the issue is that any agonist-induced changes in Ca^{2+} sensitivity will almost certainly be different in arteries that have already developed increased Ca^{2+} sensitivity, as a component of myogenic tone. While pressurized arteries develop myogenic tone, arteries mounted isometrically between wires rarely do; therefore, the operation of Ca^{2+} sensitizing systems may be different in pressurized and isometric arteries.

Evidence for the involvement of *rho*-kinase in physiological, agonist-induced contractions of pressurized rat mesenteric small arteries is the fact that the p160ROCK inhibitor, Y-27632, abolished the normal contraction, in response both to low concentrations of PE (0.1 μM) and to pressure (Van Bavel et al. 1998), but did not abolish the increase in $[Ca^{2+}]$ elicited by agonist and increased pressure. In these small arteries under near physiological conditions, it appears therefore that the activity of p160ROCK is important for setting a physiological level of Ca^{2+} sensitivity that permits changes in Ca^{2+} to activate contraction. However, this conclusion is subject to the assumption that Y-27632 is a specific inhibitor of p160ROCK.

The involvement of PKC in agonist-induced increases in Ca^{2+} sensitivity of rat mesenteric small arteries has been shown recently in an elegant study (Buus et al. 1998). This study utilized depolarized, isometrically mounted arteries that were exposed to various external $[Ca^{2+}]$, in order to control internal $[Ca^{2+}]$, in the presence or absence of noradrenaline. Internal $[Ca^{2+}]$ was measured with fura-2, and MLC_{20}-P was also measured. In the presence of noradrenaline for 40 min at an internal $[Ca^{2+}]$ of 1.0 μM, arteries developed approximately 60% more force than in the absence of NA, indicating an increase in Ca^{2+} sensitivity, since $[Ca^{2+}]$ was constant. This effect was blocked entirely by 100 nM calphostin C, implicating the involvement of PKC. (Since internal $[Ca^{2+}]$ was being measured, any effect of calphostin-C to block L-type Ca^{2+} channels, as it does in cardiac muscle, would be irrelevant.) The noradrenaline-induced Ca^{2+} sensitization also involved MP, as complete inhibition of MP abolished the ability of noradrenaline to produce Ca^{2+} sensitization. The force-MLC_{20}-P relationships were extremely steep (in both the absence and presence of noradrenaline) over a certain range, with force increasing from 30–75% of control force with an increase of MLC_{20}-P from 34–40% only (a threefold increase in internal $[Ca^{2+}]$ was required to produce this change in force). Thus, these authors suggested that additional Ca^{2+}-dependent mechanisms (i.e., in addition to MLC_{20}-P), such as the thin-filament disinhibition by phosphorylated CaD, could normally be at work to increase force.

Proposed signal events during α_1-adrenergic activation of small arteries

Given all the available evidence discussed so far, including the protein translocation studies, physiological experiments, and pharmacological data, we propose the following scheme for α_1-adrenergic activation of isobaric small arteries (Fig. 6). A video file of Ca^{2+} in a pressurized rat mesenteric small artery that shows some of the intracellular events as they actually occur is also available in the supplementary material of this article (see video file). The scheme applies specifically to bath-application of α_1-adrenoceptor agonists to arteries that develop significant myogenic tone, and which could be considered "resistance" arteries. Examples are the smaller rat mesenteric arteries and small arteries of the skeletal muscle vasculature (cremaster arteries). We consider such arteries under three conditions.

1. *Pressurized at physiological levels (e.g., 70 mm Hg) in absence of agonist* (Fig. 6a). In this situation, intracellular $[Ca^{2+}]$ will be elevated, compared to what it would be at

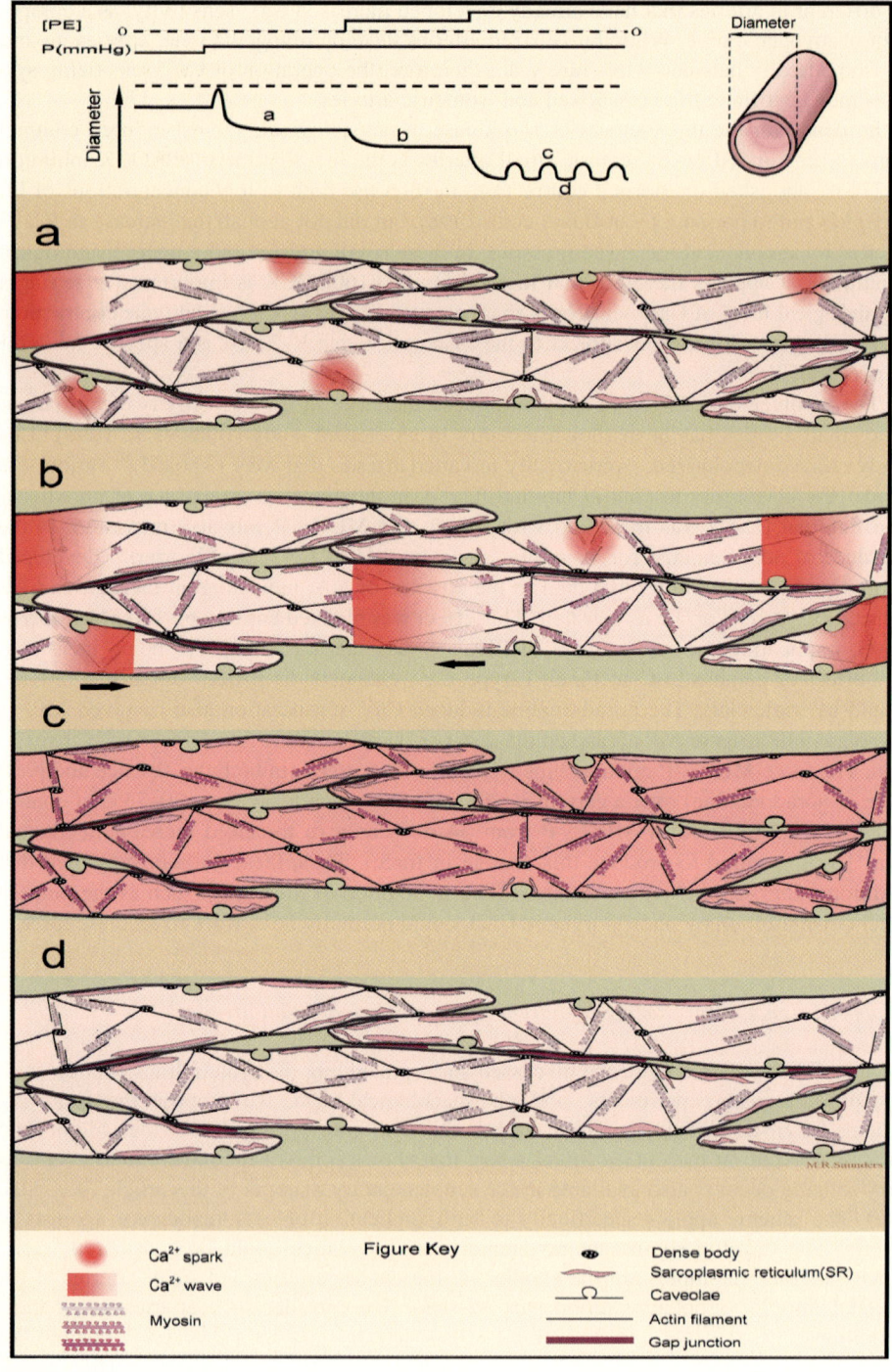

Figure Key

Ca²⁺ spark

Ca²⁺ wave

Myosin

Dense body

Sarcoplasmic reticulum(SR)

Caveolae

Actin filament

Gap junction

lower pressures. Ca^{2+} sparks will be occurring at a basal frequency and asynchronous, propagating Ca^{2+} waves will be present at a low frequency. The Ca^{2+} sensitivity of contraction is relatively high, as a result primarily of activity of the Rho-kinase pathway in inhibiting myosin phosphatase.

2. *Exposed to low levels of α_1-agonist* (Fig. 6b). In this condition, arteries constrict to a smaller diameter, without developing vasomotion. Initially, Ca^{2+} is released synchronously from the SR of all cells, primarily through InsP$_3$ sensitive Ca^{2+} channels (InsP$_3$R). Subsequent to this initial release, propagating, asynchronous Ca^{2+} waves develop. These involve primarily the InsP$_3$R. The frequency of Ca^{2+} sparks is reduced, leading to increased voltage-dependent Ca^{2+} influx. As a result of agonist activation, PKC, *rho*-A, and ERK1/2 are translocated to the cell membrane. Phosphorylation of MLC$_{20}$ increases as a result of the activation of MLCK by increased Ca^{2+}-CaM, and the PKC-dependent and ROK-dependent inhibition of myosin phosphatase (MP; with the latter constituting an increase in Ca^{2+} sensitivity, compared to the absence of agonist). Ca^{2+}-dependent thin-filament disinhibition, mediated by phosphorylated CaD is increased. Tonic vasoconstriction is maintained by InsP$_3$R-dependent Ca^{2+} waves, elevated Ca^{2+} sensitivity, and thin-filament disinhibition.

3. *Exposed to high levels of α_1-agonist* (Fig. 6c). Maximal α_1-adrenoceptor stimulation elicits a very rapid decrease in diameter, which is usually followed by vasomotion. High levels of agonist result in relatively high levels of InsP$_3$. Oscillatory vasomotion begins and continues for as long as the agonist is present, and is always associated with spatially uniform, synchronous Ca^{2+} oscillations. These oscillations in Ca^{2+} derive primarily from Ca^{2+} entering the cell through voltage-gated Ca^{2+} channels during oscillations in membrane potential. Ca^{2+} sparks are absent. Endothelium is required for the maintenance of vasomotion (at least in rat mesenteric small arteries), possibly through an influence of membrane potential of smooth muscle cells. Asynchronous propagating Ca^{2+} waves are rare. Ca^{2+} sensitivity of myofilaments is high, as a result of the agonist activation of the PKC and *rho*-kinase pathways.

Fig. 6a–d Spatio-temporal heterogeneity in Ca^{2+} signaling in adrenergically-activated small arteries. *Top*: The figure shows, in schematic form, the spatio-temporal aspects of Ca^{2+} signals within the wall of a pressurized resistance artery activated by exposure to PE. Ca^{2+} signals are heterogeneous within individual arterial smooth muscle cells, and between different cells. *Inset*: Lower trace represents arterial diameter as it changes during a step increase in pressure (*a*), and then during exposure to a relatively low [PE] (*b*) and then a higher [PE], during which vasomotion occurs (*c*, *d*). **a–d** Schematic representation of events and subcellular structures within several individual smooth muscle cells within the wall of the artery. Structures represented (see *figure key*) include dense bodies, to which intermediate filaments and myofilaments attach, SR, gap junctions between cells, and caveolae of the surface membrane. The cytoplasmic [Ca^{2+}] is represented by the Ca^{2+} sparks and Ca^{2+} waves are represented as shown in the *figure key*. Myo-filaments are shown with three levels of coloration, representing low, intermediate, and high levels of myosin light chain (LC$_{20}$) phosphorylation. Ca^{2+} sparks arise often at caveolae. SR is predominantly peripheral, and often in close apposition to caveolae. **a** In a pressurized artery, Ca^{2+} sparks are relatively frequent, and Ca^{2+} waves are relatively infrequent. LC$_{20}$ phosphorylation is relatively low. **b** At low levels of adrenergic activation, a steady vasoconstriction is achieved. The frequency of asynchronous Ca^{2+} waves is increased and the frequency of Ca^{2+} sparks decreased. At high levels of [PE], vasomotion occurs, and changes in [Ca^{2+}] are uniform and synchronous within all cells of the wall (**c**, **d**). Ca^{2+} sparks are absent, and Ca^{2+} is either uniformly high (**c**) or uniformly low (**d**), in all cells. An original video sequence which shows some of the events depicted schematically here is available in the supplementary material of this publication (see video file). The video file shows confocal images of fluo-4 fluorescence in a pressurized rat mesenteric small artery exposed to PE at 1.0 μM. Initially, the Ca^{2+} signals are asynchronous propagating Ca^{2+} waves and infrequent Ca^{2+} sparks during a period of steady vasoconstriction. Then, vasomotion develops, accompanied by synchronous Ca^{2+} oscillations

Conclusions

First, it is clear that much more than changes in Ca^{2+} are involved in adrenergically stimulated contraction of arteries. Yet, unraveling the complex interactions of the several signal transduction pathways involved remains a major challenge. Methods need to be devised by which these pathways can be studied in unperturbed cells, where the interactions can occur in a manner as close as possible to the way that they do in the intact organism. In that regard, the recent study of Miyazaki and his colleagues (Miyazaki et al. 2002) may be particularly important: the use of molecular biological techniques to insert fluorescent fusion proteins into cells, so that their translocation can be followed dynamically as an indicator (albeit imperfect) of activation. This may be particularly useful in smooth muscle, where activation involves kinase translocation. Of course, the techniques of gene transfer into intact tissues must be developed further before this can be achieved. Nevertheless, these studies hold the promise of high-resolution, dynamic imaging of physiological protein-mediated signal transduction, together with imaging Ca^{2+} signaling (as done now) within individual smooth muscle cells in the walls of intact small arteries. Of course, the use of FRET to study protein–protein interactions holds great promise for the study of signal transduction in general. Thus, we imagine a time in the future when the protein–protein interactions, protein translocations, and ion signaling may all be "seen," as it were, during physiological function.

Acknowledgments. We thank M.R. Saunders for illustrating Fig. 6 and for editing the video files included as supplementary material. Financial support for original work of W.G.W. and his coauthors was provided by NIH grant award HL64708.

Appendix

Methodology

Several methods are available for observing intracellular Ca^{2+} signaling in the walls of arteries, and each has distinct advantages and disadvantages. In fact, our understanding of the Ca^{2+} signaling mechanisms activated by adrenergic stimulation has developed in parallel with development of such new methods. Most notably, the discovery of asynchronous propagating Ca^{2+} waves in multicellular vascular smooth muscle tissue (Iino et al. 1994) had to await the confocal microscope (although agonist-induced propagating Ca^{2+} waves had been observed previously in noncontractile cultured vascular smooth muscle cells (Blatter and Wier 1992). In this section, we present very briefly the methods available for observing, with high spatial resolution, intracellular Ca^{2+} in the individual smooth muscle cells of arteries and veins, in situ (i.e., within the vascular wall). The two major problems are tissue motion (contraction), and optical working distance. Methods for mounting small arteries to measure force or observe changes in diameter have been available for some time (see Mulvany and Aalkjaer 1990). For optical measurements, in addition, the requirement for large working distance arises because the blood vessels are mounted well above the bottom of the recording chamber and because of the requirement to focus inside the vessel wall. Thus, for confocal imaging, laser light must be focused effectively through a thick aqueous layer, and fluorescence emission must be collected efficiently. In this situation, the use of objective lenses designed for aqueous specimens is essential: Oil immer-

sion objective lenses are designed for use with fixed specimens mounted in medium of re-
fractive index similar to glass, and confocal imaging in thick aqueous specimens is not
possible with them. The optical performance of a confocal microscope for imaging Ca^{2+} in
intact arteries has been critically evaluated (Miriel et al. 1999). The point-spread-function
(PSF) was measured using a single fluorescent subresolution bead placed in the interior of
a pressurized artery, providing an accurate simulation of the optical conditions during actual Ca^{2+} imaging in intact arteries. PSF derived from beads mounted on cover-glasses
would not be relevant to this type of imaging. Spatial dimensions of the PSF within the
intact arterial wall were 0.26 μm laterally, and 0.7 μm vertically. For small arteries, multi-
photon microscopy seems to provide only a modest advantage (Wier et al. 2000), probably
because the walls of small arteries are relatively thin (<20 μm). The advantage of multi-
photon microscopy would probably increase with larger arteries having thicker walls. The
essential problem in high resolution Ca^{2+} measurements is tissue motion.

Obviously, tissue motion presents a problem for imaging a constant optical section of
the arterial wall, but it is also important because sensitivity to adrenergic agonists may be
markedly dependent on wall tension (Van Bavel and Mulvany 1994), which changes when
diameter changes.

The immobilized artery

The first solution to the problem of motion during imaging Ca^{2+} with high spatial resolution was that of Iino and colleagues (Iino et al. 1994). They created an extended flat preparation by sliding a section of rat caudal artery over a flattened glass cannula. This has the
desirable effect of producing a large flat specimen that can be placed close to the bottom
of a recording chamber. For confocal optical sectioning however, even slight motion is a
problem, and in this study therefore, Cytochalasin D (5.0 μM) was added to prevent motion. This type of preparation was also used for the first recordings of Ca^{2+} sparks in multicellular vascular smooth muscle preparations (Nelson et al. 1995). While convenient for
optical recording, this preparation has two very significant disadvantages; (a) it does not
permit monitoring of contractile performance, either force or shortening, and (b) cells are
stretched an indeterminate amount, and bear an unknown amount of wall tension. The latter is important in determining sensitivity to agonists (Van Bavel and Mulvany 1994), and,
if too high, is damaging.

The confocal wire myograph

The confocal wire myograph is a development of the classical arterial ring preparation and
its use offers several advantages over flattened glass tubes. With this device, two wires
(30 μm or 40 μm in diameter) are inserted into the lumen of the artery, and the artery is
then stretched into a flattened oval shape. One of the wires is fixed, and the other is attached to a force transducer. The wires are mounted on micromanipulators such that the
artery can be lowered to just above the chamber bottom, which must be a glass cover slip.
The artery must be lowered to within the working distance of a "water" objective lens
(typically 220 μm). A major advantage of this device is that arteries can be stretched to
provide a known (and reproducible) wall tension, and, of course, force development can
be measured. The smooth muscle cells are stretched out flat, and thus this method also
provides a good specimen for optical imaging. Its primary disadvantage is that the artery

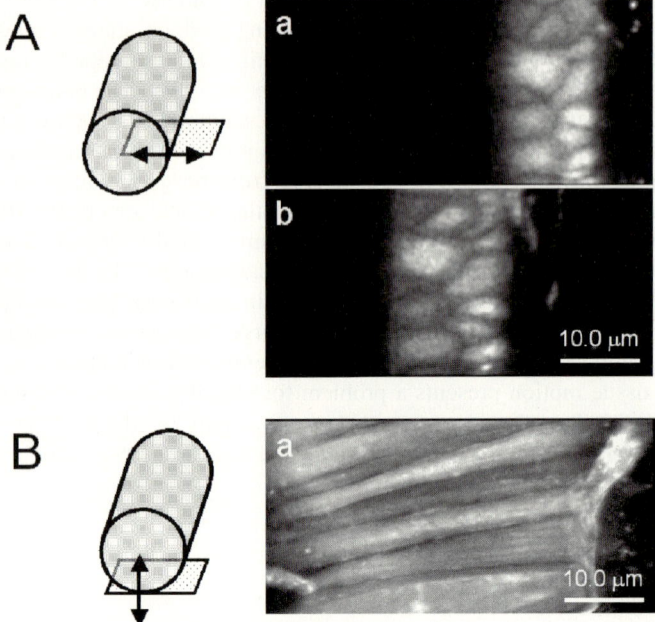

Fig. 7A, B Confocal optical sectioning of the vascular wall of pressurized mesenteric arteries during vaso-motion **A** and steady vasoconstriction **B**. **A**, *inset*) The stippled plane indicates a "horizontal radial" optical section as obtained by the confocal laser-scanning microscope. Individual smooth muscle cells (*SMC*) will appear in crosssection in this type of image **A**, *a*, *b*). This optical section is coplanar with a plane through the geometrical center of the artery. Note that the vertical component of vasomotion (*arrow*) is zero in this case, thus obviating movement artifacts that occur when the confocal microscope is focused in other planes. Individual smooth muscle cells (*SMC*) can be observed unambiguously during the vasomotion **A**, *a*, *b*), al-though their shape changes. **B** *Inset*: The stippled plane indicates a "tangential" optical section. High reso-lution horizontal tangential optical sections, showing the individual SMC in longitudinal section (**B**, *a*). This type of optical section can be used during steady vasoconstriction and in arteries mounted isometrically, on glass cannulae. (From Mauban et al. 2001)

is forced to contract isometrically, something that probably does not occur physiologically. In this preparation, wall tension increases during contraction.

The pressurized artery

This type of preparation clearly is the most physiological. Through the use of pressure transducers and a servo-controlled pump, intraluminal pressure or flow can be controlled. Arteries can be made to contract at constant pressure (isobaric). Provided the pressure is not too great, activation will then result in a decrease in arterial radius (R), which can be readily measured with video-based edge detectors. In this case, (according to the Law of LaPlace, $T = P \times R$) wall tension actually decreases (since internal pressure, P, remains con-stant). Of course, the major disadvantage of this method is that this unrestricted motion interferes with confocal optical sectioning. In fact, the first confocal images of intracellu-lar Ca^{2+} in pressurized arteries (Miriel et al. 1999) were obtained only in arteries at rest or in arteries that had constricted in response to adrenergic agonists, but which were not undergoing vasomotion. By focusing the confocal microscope in a plane through the center of the artery, however, some of the effects of wall motion can be lessened (Fig. 7, from Mauban et al. 2001). In this plane of focus, the smooth muscle cells of the arterial wall are viewed in crosssection, and the same region of a particular cell stays roughly in focus, despite motion, because the motion in that one plane is entirely horizontal (there is no vertical component of motion in the arterial wall in the one plane that goes through the center of the artery). Of course, if the artery is viewed in an optical section through the wall at the bottom, then the opposite is true (all the motion is vertical), and the specimen disappears from view quickly. Through the use of an appropriate feedback system, the pressurized artery can also be made to contract truly isometrically (Van Bavel and Mulvany 1994). These authors devised a system in which changes of the vessel diameter resulted in a change in the amount of fluorescence light from a dye contained in the vessel lumen. Any change in diameter from the desired value was sensed by this system and was used in a feedback circuit to clamp arterial diameter at the desired level (by adjusting internal pressure). This system yielded important information about the effect of wall tension on agonist sensitivity. Such a system has not yet been combined with confocal microscopy of Ca^{2+}-indicator signals.

Supplementary material

The data on Ca^{2+} signaling are best viewed dynamically, and several video files are included as supplementary material with this article. Video files viewable with standard computer media players have also been published as supplementary data with several of the studies reviewed here (Zang et al. 2001; Mauban et al. 2001; Zhang et al. 2002; Lamont and Wier 2002).

References

Abe J, Baines CP, Berk BC (2000) Role of mitogen-activated protein kinases in ischemia and reperfusion injury: the good and the bad. Circ Res 86:607–609

Abraham ST, Benscoter H, Schworer CM, Singer HA (1997) A role for Ca^{2+}/calmoduin-dependent protein kinase II in the mitogen-activated protein kinase signaling cascade of cultured rat aortic vascular smooth muscle cells. Circ Res 81:575–584

Aburto T, Jinsi A, Zhu Q, Deth RC (1995) Involvement of protein kinase C activation in α_2-adrenoceptor-mediated contractions of rabbit saphenous vein. Eur J Pharmacol 277:35–44

Adam LP, Graceffa P, Haeberle JR (1997) Caldesmon's effects on actin filament motility, in vitro, are reversed by phosphorylation with MAPK (abstract). Biophys J 72:A176:216

Albert AP and Large WA (2002) Activation of store-operated channels by noradrenaline via protein kinase C in rabbit portal vein myocytes. J Physiol 544:113–125.

Alessi D, MacDougall LK, Sola MM, Ikebe M, Cohen P (1992) The control of protein phosphatase-1 by targeting subunits. The major myosin phosphatase in avian smooth muscle is a novel form of protein phosphatase-1. Eur J Biochem 210:1023–1035

Amano M, Ito M, Kimura K, Fukata Y, Chihara K, Nakano T, Matsuura Y, Kaibuchi K (1996) Phosphorylation and activation of myosin by Rho-associated kinase (Rho-kinase). J Biol Chem 271:20246–20249

Arnon A, Hamlyn JM, Blaustein MP (2000) Na^+ entry via store-operates channels modulates Ca^{2+} signalling in arterial myocytes. Am J Physiol 278:C163–C173

Aromolaran AS, Albert AP and Large WA (2000) Evidence for myosin light chain kinase mediating noradrenaline-evoked cation current in rabbit portal vein myocytes. J Physiol 524:853–863

Asada Y, Yamazawa T, Hirose K, Takasaka T, Iino M (1999) Dynamic Ca^{2+} signalling rat arterial smooth muscle cells under the control of local renin-angiotensin system. J Physiol 521:497–505

Ashida T, Schaeffer J, Goldman WF, Wade JB, Blaustein MP (1988) Role of SR in arterial contraction: comparison of ryanodine's effect in a conduit and a muscular artery. Circ Res 62:854–863

Bao JX, Stjarne L (1993) Dual contractile effects of ATP released by field stimulation revealed by effects of α,β-methylene ATP and suramin in rat tail artery. Br J Pharmacol 110:1421–1428

Baro I, Eisner DA (1995) Factors controlling changes in intracellular Ca^{2+} concentration produced by noradrenalinee in rat mesenteric artery smooth muscle cells. J Physiol 482:247–258

Baro I, Eisner D (1992) The effects of thapsigargin on $[Ca^{2+}]_i$ in isolated rat mesteric artery vascular smooth muscle cells. Pflug Arch 420:115–117

Baro L, O'Neill SC, Eisner DA (1993) Changes of intracellular $[Ca^{2+}]$ during refilling of SR in rat ventricular and vascular smooth muscle. J Physiol 465:21–41

Batchelor TJ, Sadaba JR, Ishola A, Pacaud P, Munsch CM, Beech DJ (2001) Rho-kinase inhibitors prevent agonist-induced vasospasm in human internal mammary artery. Br J Pharmacol 132:302–308

Bayley PM, Findlay WA, Martin SR (1996) Target recognition by calmodulin: dissecting the kinetics and affinity of interaction using short peptide sequences. Protein Sci 5:1215–1228

Beech DJ (2002) SOCs—store-operated channels in vascular smooth muscle? J Physiol 544:1

Benham CD, Bolton TB (1986) Spontaneous transient outward currents in single visceral and vascular smooth muscle cells of the rabbit. J Physiol 499:291–306

Blatter LA, Wier WG (1992) Agonist-induced $[Ca^{2+}]_i$ waves and Ca^{2+}-induced Ca^{2+} release in mammalian vascular smooth muscle cells. Am J Physiol 263:H576–H586

Blaustein MP, Lederer WJ (1999) Sodium/calcium exchange: Its physiological implications. Physiol Revs 79:763–854

Blaustein MP, Golovina VA, Song H, Choate J, Lencesova L, Robinson SW, Wier WG (2002) Organization of Ca^{2+} stores in vascular smooth muscle: functional implications. Novartis Found Symp 246:125–137; discussion 137–141, 221–227

Bogatcheva NV, Gusev NB (1995) Interaction of smooth muscle calponin with phospholipids. FEBS Lett 371:123–126

Boittin F-X, Macrez N, Halet G, Mironneau J (1999) Norepinephrine-induced Ca^{2+} waves depend on InsP$_3$ and ryanodine receptor activation in vascular myocytes. Am J Physiol Cell Physiol 277:C139–C151

Bolz SS, Galle J, Derwand R, de Wit C, Pohl U (2000) Oxidized LDL increases the sensitivity of the contractile apparatus in isolated resistance arteries for Ca^{2+} via a rho- and rho kinase-dependent mechanism. Circulation 102:2402–2410

Bonev A, Jaggar JH, Rubart M, Nelson MT (1997) Activators of protein kinase C decrease Ca^{2+} spark frequency in smooth muscle cells from cerebral arteries. Am J Physiol 273:C2090–C2095

Bradley AB, Morgan KG (1987) Alteration in cytoplasmic calcium sensitivity during porcine coronary artery contractions as detected by aequorin. J Physiol 385:437–448

Braun AP, Schulman H (1995) The multifunctional calcium/calmodulin-dependent protein kinase: from form to function. Annu Rev Physiol 57:417–445

Brenner R, Perez GJ, Bonev AD, Eckman DM, Kosek JC, Wiler SW, Patterson AJ, Nelson MT, Aldrich RW (2000) Vasoregulation by the $\beta1$ subunit of the calcium-activated potassium channel. Nature 407:870–876

Brocke L, Srinivasan M, Schulman H (1995) Developmental and regional expression of multifunctional Ca^{2+}/calmodulin-dependent protein kinase isoforms in rat brain. J Neurosci 15:6797–6808

Brocke L, Chiang LW, Wagner PD, Schulman H (1999) Functional implications of the subunit composition of neuronal CaM kinase II. J Biol Chem 274:22713–22722

Brozovich FV (1995) PKC regulates agonist-induced force enhancement in single α-toxin-permeabilized vascular smooth muscle cells. Am J Physiol 268(5 Pt 1):C1202–C1206

Buus CL, Aalkjaer C, Nilsson H, Juul B, Moller JV, Mulvany MJ (1998) Mechanisms of Ca^{2+} sensitization of force production by noradrenaline in rat mesenteric small arteries. J Physiol 510:590

Cauvin C, Tejerina M, Hwang O, Kai-Yamamoto M, Van Breemen C (1988) The effects of Ca^{2+} antagonists on isolated rat and rabbit mesenteric resistance vessels. What determines the sensitivity of agonist-activated vessels to Ca^{2+} antagonists? Ann N Y Acad Sci 522:338–350

Chalovich JM (1988) Caldesmon and thin-filament regulation of muscle contraction. Cell Biophys 12:73–85

Chatterjee M, Tejada M (1986) Phorbol ester-induced contraction in chemically skinned vascular smooth muscle. Am J Physiol 251:C356–C61

Chin D, Means AR (2000) Calmodulin: a prototypical calcium sensor. Trends Cell Biol 10:322–328

Colbran RJ, Soderling TR (1990) Calcium/calmodulin-dependent protein kinase II. Curr Top Cell Regul 31:181–221

Cook AK, Carty M, Singer CA, Yamboliev IA, Gerthoffer WT (2000) Coupling of M_2 muscarinic receptors to ERK MAP kinases and caldesmon phosphorylation in colonic smooth muscle. Am J Physiol Gastrointest Liver Physiol 278:G429–G437

Coussin F, Macrez N, Morel JL, Mironneau J (2000) Requirement of ryanodine receptor subtypes 1 and 2 for Ca^{2+}-induced Ca^{2+} release in vascular myocytes. J Biol Chem 275:9596–9603

Crowley CM, Lee CH, Gin SA, Keep AM, Cook RC, Van Breemen C (2002) The mechanism of excitation-contraction coupling in PE-stimulated human saphenous vein. Am J Physiol Heart Circ Physiol 283:H1271–H1281

Curtis TM and Scholfield CN (2001) Nifedipine blocks Ca^{2+} store refilling through a pathway not involving L-type Ca^{2+} channels in rabbit arteriolar smooth muscle. J Physiol 532:609–623

D'Angelo G, Adam LP (2002) Inhibition of ERK attenuates force development by lowering myosin light chain phosphorylation. Am J Physiol Heart Circ Physiol 282:H602–H610

Danthuluri NR, Deth RC (1984) Phorbol ester-induced contraction of arterial smooth muscle and inhibition of α-adrenergic response. Biochem Biophys Res Commun 125:1103–1109

Deisseroth K, Heist EK, Tsien RW (1998) Translocation of calmodulin to the nucleus supports CREB phosphorylation in hippocampal neurons. Nature 392:198–202

De Koninck P, Schulman H (1998) Sensitivity of CaM kinase II to the frequency of Ca^{2+} oscillations. Science 279:227–230

Deng JT, Van Lierop JE, Sutherland C, Walsh MP (2001) Ca^{2+}-independent smooth muscle contraction. a novel function for integrin-linked kinase. J Biol Chem 276:16365–16373

Dessy C, Kim I, Sougnez CL, Laporte R, Morgan KG (1998) A role for MAP kinase in differentiated smooth muscle contraction evoked by α-adrenoceptor stimulation. Am J Physiol Cell 275:C1081–C1086

Dillon PF, Aksoy MO, Driska SP, Murphy RA (1981) Myosin phosphorylation and the crossbridge cycle in arterial smooth muscle. Science 211:495–497

Diver JM, Sage SO, Rosado JA (2001) The inositol trisphosphate receptor antagonist 2- aminoethoxydiphenylborate (2-APB) blocks Ca^{2+} entry channels in human platelets: cautions for its use in studying Ca^{2+} influx. Cell Calcium 30:323–329

Dreja K, Nordstrom I, Hellstrand P (2001) Rat arterial smooth muscle devoid of ryanodine receptor function: effects on cellular Ca^{2+} handling. Br J Pharmacol 132:1957–1966

Dunn AR, Mann GB, Fowler KJ, Grail D, Hibbs ML, Alexander WS, Walker F, Burgess AW (1994) Insights into the physiology of $TGF\alpha$ and signaling through the EGF receptor revealed by gene targeting and acts of nature. Princess Takamatsu Symp 24:276–289

Earley JJ, Su X, Moreland RS (1998) Caldesmon inhibits active crossbridges in unstimulated vascular smooth muscle: an antisense oligodeoxynucleotide approach. Circ Res 83:661–667

Eto M, Ohmori T, Suzuki M, Furuya K, Morita F (1995) A novel protein phosphatase-1 inhibitory protein potentiated by protein kinase C. Isolation from porcine aorta media and characterization. J Biochem (Tokyo) 118:1104–1107

Eto M, Senba S, Morita F, Yazawa M (1997) Molecular cloning of a novel phosphorylation-dependent inhibitory protein of protein phosphatase-1 (CPI17) in smooth muscle: its specific localization in smooth muscle. FEBS Lett 410:356–360

Feng J, Ito M, Ichikawa K, Isaka N, Nishikawa M, Hartshorne DJ, Nakano T (1999) Inhibitory phosphorylation site for rho-associated kinase on smooth muscle myosin phosphatase. J Biol Chem 274:37385–37389

Fleckenstein-Grun G (1996) Calcium antagonism in vascular smooth muscle cells. Pflugers Arch 432:R53–R60

Flemming R, Cheong A, Dedman AM and Beech DJ (2002) Discrete store-operated calcium influx into an intracellular compartment in rabbit arteriolar smooth muscle. J Physiol 543:455–464

Flynn ER, Bradley KN, Muir TC, McCarron JG (2001) Functionally separate intracellular Ca^{2+} stores in smooth muscle. J Biol Chem 276:36411–36418

Garcha RS, Hughes AD (1995) Inhibition of norepinephrine and caffeine-induced activation by ryanodine and thapsigargin in rat mesenteric arteries. J Cardiovasc Pharmacol 25:840–846

Gerthoffer WT, Murphy KA, Gunst SJ (1989) Aequorin luminescence, myosin phosphorylation, and active stress tracheal smooth muscle. Am J Physiol 257:C1062–C1068

Gerthoffer WT, Yamboliev IA, Shearer M, Pohl J, Haynes R, Dang S, Sato K, Sellers JR (1996) Activation of MAP kinases and phosphorylation of caldesmon in canine colonic smooth muscle. J Physiol 495.3:597–609

Gimona M, Mital R (1998) The single CH domain of calponin is neither sufficient nor necessary for F-actin binding. J Cell Sci 111:1813–1821

Gimona M, Winder SJ (1998) Single calponin homology domains are not actin binding domains. Curr Biol 8:R674–R675

Gimona M, Small JV (1996) Calponin. In Bárány M (ed) Biochemistry of smooth muscle. Academic Press, San Diego, pp 91–104

Gimona M, Herzog M, Vandekerckhove J, Small JV (1990) Smooth muscle specific expression of calponin. FEBS Lett 274:159–162

Gitterman DP, Evans RJ (2001) Nerve-evoked P2X receptor contractions of rat mesenteric arteries; dependence on vessel size and lack of role of L-type calcium channels and calcium-induced calcium release. Br J Pharmacol 132:1201–1208

Goeckeler ZM, Masaracchia RA, Zeng Q, Chew T-L, Gallagher P, Wysolmerski RB (2000) Phosphorylation of myosin light chain kinase by p21-activated kinase PAK2. J Biol Chem 275:18366–18374

Gokina NI, Osol G (1998) Temperature and protein kinase C modulate myofilament Ca^{2+} sensitivity in pressurized rat cerebral arteries. Am J Physiol 274:H1920–H1927

Gokina NI, Knot HJ, Nelson MT (1999) Increased Ca^{2+}sensitivity as a key mechanism of PKC-induced constriction in pressurized cerebral arteries. Am J Physiol 46:H1178–H1188

Golovina VA, Blaustein MP (1997) Spatially and functionally distinct Ca^{2+} stores in sarcoplasmic and endoplasmic reticulum. Science 275:1643–1648

Gong MC, Fujihara H, Somlyo AV, Somlyo AP (1997) Translocation of rhoA associated with Ca^{2+} sensitization of smooth muscle. J Biol Chem 272:10704–10709

Gordienko DV, Bolton TB (2002) Crosstalk between ryanodine receptors and $IP_{(3)}$ receptors as a factor shaping spontaneous $Ca^{(2+)}$-release events in rabbit portal vein myocytes. J Physiol 542:743–762

Gordienko DV, Bolton, TB, Cannell, MB (1998) Variability in spontaneous subcellular calcium release in guinea-pig ileum smooth muscle cells. J Physiol 507:707–720

Gordienko DV, Greenwood IA, Bolton TB (2001) Direct visualization of sarcoplasmic reticulum regions discharging Ca^{2+} sparks in vascular myocytes. Cell Calcium 29:13–28

Gorenne I, Su X, Moreland RS (1998) Inhibition of p42 and p44 MAP kinase does not alter smooth muscle contraction in swine carotid artery. Am J Physiol 44:H131–H138

Greenwood IA, Ledoux J, Leblanc N (2001) Differential regulation of Ca^{2+}-activated Cl^- currents in rabbit arterial and portal vein smooth muscle cells by Ca^{2+}-calmodulin-dependent kinase. J Physiol 534(Pt. 2):395–408

Grover AK, Xu A, Samson SE, Narayanan N (1996) SR Ca^{2+} pump in pig coronary artery smooth muscle is regulated by a novel pathway. Am J Physiol 271:C181–C7

Gunst SJ, Tang DD (2000) The contractile apparatus and mechanical properties of airway smooth muscle. Eur Respir J 15:600–616

Gunst SJ, Gerthoffer WT, al-Hassani MH (1992) Ca^{2+} sensitivity of contractile activation during muscarinic stimulation of tracheal muscle. Am J Physiol Cell Physiol 263:C1258–C1265

Gunst SJ, al-Hassani MH, Adam LP (1994) Regulation of isotonic shortening velocity by second messengers in tracheal smooth muscle. Am J Physiol Cell 266:C684–C691

Gustafsson H, Bulow A, Nilsson H (1994) Rhythmic contractions of isolated pressurized small arteries from rat. Acta Physiol Scand 152:145–152

Haeberle JR (1999) Thin-filament linked regulation of smooth muscle myosin. J Muscle Res Cell Motil 20:363–370

Hai C-M, Murphy RA (1988) Crossbridge phosphorylation and regulation of latch state in smooth muscle. Am J Physiol 254:C99–C106

Hai C-M, Murphy RA (1989) Ca^{2+}, crossbridge phosphorylation, and contraction. Annu Rev Physiol 51:285–298

Halayko AJ, Solway J (2001) Molecular mechanisms of phenotypic plasticity in smooth muscle cells. J Appl Physiol 90:358–368

Hamaguchi T, Ito M, Feng J, Seko T, Koyama M, Machida H, Takase K, Amano M, Kaibuchi K, Hartshorne DJ, Nakano T (2000) Phosphorylation of CPI-17, an inhibitor of myosin phosphatase, by protein kinase N. Biochem Biophys Res Commun 274:825–830

Hartshorne DJ, Ito M, Erdodi F (1998) Myosin light chain phosphatase: subunit composition, interactions and regulation. J Muscle Res Cell Motil 19:325–341

Hedges JC, Oxhorn BC, Carty M, Adam LP, Yamboliev IA, Gerthoffer WT (2000) Phosphorylation of caldesmon by ERK MAP kinases in smooth muscle. Am J Physiol Cell 278:C718–C726

Hemric ME, Tracy PB, Haeberle JR (1994) Caldesmon enhances the binding of myosin to the cytoskeleton during platelet activation. J Biol Chem 269:4125–4128

Himpens B, Matthijs G, Somlyo AV, Butler TM, Somlyo AP (1988) Cytoplasmic free calcium, myosin light chain phosphorylation, and force in phasic and tonic smooth muscle. J Gen Physiol 92:713–729

Hirst GD, Edwards FR (1989) Sympathetic neuroeffector transmission in arteries and arterioles. Physiol Rev 69:546–604

Horowitz A, Clement-Chomienne O, Walsh MP, Morgan KG (1996a) ε-Isoenzyme of protein kinase C induces a Ca^{2+}-independent contraction in vascular smooth muscle. Am J Physiol 271:C589–C594

Horowitz A, Menice CB, Laporte R, Morgan KG (1996b) Mechanisms of smooth muscle contraction. Physiol Rev 76:967–1003

Hoth M, Penner R (1993) Calcium release-activated calcium current in rat mast cells. J Physiol 465:359–386

Hulvershorn J, Gallant C, Wang C-LW, Dessy C, Morgan KG (2001) Calmodulin levels are dynamically regulated in living vascular smooth muscle cells. Am J Physiol Heart Physiol 280:H1422–H1426

Hwang KS, Van Breemen C (1987) Ryanodine modulation of ^{45}Ca efflux and tension in rabbit aortic smooth muscle. Pflug Arch 408:343–350

Ichikawa K, Ito M, Hartshorne DJ (1996a) Phosphorylation of the large subunit of myosin phosphatase and inhibition of phosphatase activity. J Biol Chem 271:4733–4740

Ichikawa K, Hirano K, Ito M, Tanaka J, Nakano T, Hartshorne DJ (1996b) Interactions and properties of smooth muscle myosin phosphatase. Biochemistry 35:6313–6320

Iino M (2002) Family affairs of intracellular Ca^{2+}-release channels. J Physiol 542:667

Iino M, Kasai H, Yamazawa T (1994) Visualization of neural control on intracellular Ca^{2+} concentration in single vascular smooth muscle cells in situ. EMBO J 13:5026–5031

Ikebe M, Reardon S (1988) Binding of caldesmon to smooth muscle myosin. J Biol Chem 263:3055–3058

Jaggar JH (2001) Intravascular pressure regulates local and global Ca^{2+} signaling in cerebral artery smooth muscle cells. Am J Physiol 281:C439–C448

Jaggar JH, Nelson MT (2000) Differential regulation of Ca^{2+} sparks and Ca^{2+} waves by UTP in rat cerebral artery smooth muscle cells. Am J Physiol Cell Physiol 279:C1528–C1539

Jaggar JH, Porter VA, Lederer WJ, Nelson MT (2000) Calcium sparks in smooth muscle. Am J Physiol 278:C235–C256

Jaggar JH, Stevenson AS, Nelson MT (1998a) Voltage dependence of Ca^{2+} sparks in intact cerebral arteries. Am J Physiol 274:C1755–1761

Jaggar JH, Wellman GC, Heppner TJ, Porter VA, Perez GJ, Gollasch M, Kleppisch T, Stevenson AS, Lederer WJ, Knot HJ, Bonev AD, Nelson MT (1998b) Ca^{2+} channels, ryanodine receptors and Ca^{2+}-activated K^+ channels: a functional unit for regulating arterial tone. Acta Physiol Scand 164:577–587

Janiak R, Wilson SM, Montague S, Hume JR (2001) Heterogeneity of calcium stores and elementary release events in canine pulmonary arterial smooth muscle cells. Am J Physiol 280:C22–C33

Je HD, Gangopadhyay SS, Ashworth TD, Morgan KG (2001) Calponin is required for agonist-induced signal transduction—evidence from an antisense approach in ferret smooth muscle. J Physiol 537(Pt 2):567–577

Jiang MJ, Morgan KG (1987) Intracellular calcium levels in phorbol ester-induced contractions of vascular muscle. Am J Physiol 253:H1365–H1371

Jiang MJ, Morgan KG (1989) Agonist-specific myosin phosphorylation and intracellular calcium during isometric contractions of arterial smooth muscle. Pflug Archiv 413:637–643

Johnson D, Cohen P, Chen MX, Chen YH, Cohen PT (1997) Identification of the regions on the M110 subunit of protein phosphatase 1 M that interact with the M21 subunit and with myosin. Eur J Biochem 244:931–939

Johnson JD, Snyder C, Walsh M, Flynn M (1996) Effects of myosin light chain kinase and peptides on Ca^{2+} exchange with the N- and C-terminal Ca^{2+} binding sites of calmodulin. J Biol Chem 271:761–767

Julou-Schaeffer G, Freslon JL (1988) Effect of ryanodine on tension development in rat aorta and mesenteric resistance arteries. Br J Pharmacol 95:605–613

Kamm KE, Stull JT (1985) The function of myosin and myosin light chain kinase phosphorylation in smooth muscle. Ann Rev Pharmacol Toxicol 25:593–620

Kamm KE, Stull JT (1989) Regulation of smooth muscle contractile elements by second messengers. Annu Rev Physiol 51:299–313

Kaneko T, Amano M, Maeda A, Goto H, Takahashi K, Ito M, Kaibuchi K (2000) Identification of calponin as a novel substrate of Rho-kinase. Biochem Biophys Res Commun 273:110–116

Kanmura Y, Missiaen L, Raeymaekers L, Casteels R (1988) Ryanodine reduces the amount of calcium in intracellular stores of smooth-muscle cells of the rabbit ear artery. Pflugers Arch 413:153–159

Karaki H, Ozaki H, Hori M, Mitsui-Saito M, Amano K, Harada K, Miyamoto S, Nakazawa H, Won KJ, Sato K (1997) Calcium movements, distribution, and functions in smooth muscle. Pharmacol Rev 49:157–230

Kasai Y, Yamazawa T, Sakurai T, Taketani Y, Iino M (1997) Endothelium-dependent frequency modulation of Ca^{2+} signalling in individual vascular smooth muscle cells of the rat. J Physiol 504:349–357

Katsuyama H, Wang C-LA, Morgan KG (1992) Regulation of vascular smooth muscle tone by caldesmon. J Biol Chem 267:14555–14558

Khalil RA, Morgan KG (1993) PKC-mediated redistribution of mitogen-activated protein kinase during smooth muscle activation. Am J Physiol 265:C406–C411

Khalil RA, Lajoie C, Morgan KG (1994) In situ determination of the $[Ca^{2+}]_i$ threshold for translocation of the α protein kinase C isoform. Am J Physiol: Cell 266:C1544–C1551

Khalil RA, Menice CB, Wang C-LA, Morgan KG (1995) Phosphotyrosine-dependent targeting of mitogen-activated protein kinase in differentiated contractile vascular cells. Circ Res 76:1101–1108

Kim I, Je H-D, Gallant C, Zhan Q, Van Riper D, Badwey JA, Singer HA, Morgan KG (2000) Ca^{2+}-calmodulin-dependent protein kinase II-dependent activation of contractility in ferret aorta. J Physiol 526:367–374

Kimura K, Ito M, Amano M, Chihara K, Fukata Y, Nakafuku M, Yamamori B, Feng J, Nakano T, Okawa K, Iwamatsu A, Kaibuchi K (1996) Regulation of myosin phosphatase by Rho and Rho-associated kinase (Rho-kinase). Science 273:245–248

Kitazawa T, Takizawa N, Ikebe M, Eto M (1999) Reconstitution of protein kinase C-induced contractile Ca^{2+} sensitization in triton X-100-demembranated rabbit arterial smooth muscle. J Physiol 520(Pt 1):139–152

Kitazawa T, Eto M, Woodsome TP, Brautigan DL (2000) Agonists trigger G protein-mediated activation of the CPI-17 inhibitor phosphoprotein of myosin light chain phosphatase to enhance vascular smooth muscle contractility. J Biol Chem 275:9897–9900

Knot HJ (2001) Calcium sparks unleashed in vascular smooth muscle: lessons from the RyR_3 knockout mouse. Circ Res 89:941–943

Kureishi Y, Kobayashi S, Amano M, Kimura K, Kanaide H, Nakano T, Kaibuchi K, Ito M (1997) Rho-associated kinase directly induces smooth muscle contraction through myosin light chain phosphorylation. J Biol Chem 272:12257–12260

Lagaud GJ, Skarsgard PL, Laher I, Van Breemen C (1999) Heterogeneity of endothelium-dependent vasodilation in pressurized cerebral and small mesenteric resistance arteries of the rat. J Pharmacol Exp Ther 290:832–839

Lamont C, Wier WG (2002) Evoked and spontaneous purinergic junctional Ca^{2+} transients in rat small arteries. Circ Res 91:4454–4456

Lee C-H, Poburko D, Sahota P, Sandhu J, Ruehlmann DO, Van Breeman C (2001) The mechanism of phenylephrine-mediated $[Ca^{2+}]_i$ oscillations underlying tonic contraction in the rabbit IVC. J Physiol 534.3:641–650

Lee C-H, Poburko D, Kuo K-H, Seow CY, Van Breeman C (2002a) Ca^{2+} oscillations, gradients, and homeostasis in vascular smooth muscle. Am J Physiol Heart Circ Physiol 282:H1571–H1583

Lee C-H, Rahimian R, Szado T, Sandhu J, Poburko D, Behra T, Chan L, Van Breeman C (2002b) Requirement for the opening of IP_3-sensitive Ca^{2+} channels and SOC in α_1-adrenergic receptor-mediated constriction of the rabbit IVC. Am J Physiol Heart Circ Physiol (in press)

Lee Y-H, Kim I, Laporte R, Walsh MP, Morgan KG (1999) Isozyme-specific inhibitors of PKC translocation: effects on contractility of single permeabilized vascular muscle cells of the ferret. J Physiol (Lond) 517:709–720

Lee Y-H, Gallant C, Guo H, Li Y, Wang C-LA, Morgan KG (2000) Regulation of vascular smooth muscle tone by N-terminal region of caldesmon: possible role of tethering actin to myosin. J Biol Chem 275:3213–3220

Leinweber B, Parissenti AM, Gallant C, Gangopadhyay SS, Kirwan-Rhude A, Leavis PC, Morgan KG (2000) Regulation of protein kinase C by the cytoskeletal protein calponin. J Biol Chem 275:40329–40336

Leinweber BD, Leavis PC, Grabarek Z, Wang C-LA, Morgan KG (1999) Extracellular regulated kinase (ERK) interaction with actin and the calponin homology (CH) domain of actin binding proteins. Biochem J 344:117–123

Lesh RE, Nixon GR, Fleischer S, Airey JA, Somlyo AP, Somlyo AV (1998) Localization of ryanodine receptors in smooth muscle. Circ Res 82:175–185

Li L, Guo H, Wang C-LA (2001) Effect of ERK-phosphorylation on actin binding of caldesmon (abstract). Biophys J 80:359a

Li Y, Zhuang S, Guo H, Mabuchi K, Lu RC, Wang CA (2000) The major myosin-binding site of caldesmon resides near its N-terminal extreme. J Biol Chem 275:10989–10994

Lin P, Luby-Phelps K, Stull JT (1999) Properties of filament-bound myosin light chain kinase. J Biol Chem 274:5987–5994

Lohn M, Furstenau M, Sagach V, Elger M, Schulze W, Luft FC, Haller H, Gollasch M (2000) Ignition of calcium sparks in arterial and cardiac muscle through caveolae. Circ Res 87:1034–1039

Lohn M, Jessner W, Furstenau M, Wellner M, Sorrentino V, Haller H, Luft FC, Gollasch M (2001) Regulation of calcium sparks and spontaneous transient outward currents by RyR$_3$ in arterial vascular smooth muscle cells. Circ Res 89:1051–1057

Lopez-Lopez JR, Shacklock PS, Balke CW, Wier WG (1995) Local calcium transients triggered by single L-type calcium channel currents in cardiac cells. Science 268:1042–1045

Luby-Phelps K, Hori M, Phelps JM, Won D (1995) Ca^{2+}-regulated dynamic compartmentalization of calmodulin in living smooth muscle cells. J Biol Chem 270:21532–21538

Mabuchi K, Li B, Ip W, Tao T (1997) Association of calponin with desmin intermediate filaments. J Biol Chem 272:22662–22666

MacDonald JA, Borman MA, Muranyi A, Somlyo AV, Hartshorne DJ, Haystead TA (2001) Identification of the endogenous smooth muscle myosin phosphatase-associated kinase. Proc Natl Acad Sci USA 98:2419–2424

Marston S, Levine BA, Gao Y, Evans J, Patchell VB, El-Mezgueldi M, Fattoum A, Vorotnikov AV (2001) MAP kinase phosphorylation at serine 702 alters structural and actin binding properties of caldesmon (abstract). Biophys J 80:69a

Marston SB, Smith CWJ (1984) Purification and properties of Ca^{2+}-regulated thin filaments and F-actin from sheep aorta smooth muscle J Muscle Res Cell Motility 5:559–575

Marston SB, Trevett RM, Walters M (1980) Calcium ion-regulated thin filaments from vascular smooth muscle. Biochem J 185:355–365

Matrougui K, Tanko LB, Loufrani L, Gorny D, Levy BI, Tedgui A, Henrion D (2001) Involvement of Rho-kinase and the actin filament network in angiotensin II-induced contraction and extracellular signal-regulated kinase activity in intact rat mesenteric resistance arteries. Arterioscler Thromb Vasc Biol 21:1288–1293

Matthew JD, Khromov AS, Somlyo AV, Somlyo AP, Karaki H, Tsuchiya T, Takahashi K (2000) Ca^{2+}-sensitization of smooth muscle in calponin knockout mouse (abstract). Biophys J 78:647

Mauban J, Lamont C, Balke CW, Wier WG (2001) Adrenergic stimulation of rat resistance arteries affects Ca^{2+} sparks, Ca^{2+} waves, and Ca^{2+} oscillations. Am J Physiol 280:H2399–H2405

McCaron JG, McGeown JG, Reardon S, Ikebe M, S FF, Walsh JV (1992) Calcium-dependent enhancement of calcium current in smooth muscle by calmodulin-dependent protein kinase II. Nature 357:74–77

McGeown JG, McCarron JG, Drummond RM, Fay FS (1998) Calcium-calmodulin-dependent mechanisms accelerate calcium decay in gastric myocytes from Bufo marinus. J Physiol 506(Pt 1):95–107

Meisheri KD, Ruegg JC, Paul RJ (1985) Studies on skinned fiber preparations. In: Grover AK, Daniel EE (eds) Calcium and contractility. Humana Press, Clifton, NJ, pp 191–224

Menice CB, Hulvershorn J, Adam LP, Wang C-LA, Morgan KG (1997) Calponin and mitogen-activated protein kinase signaling in differentiated vascular smooth muscle. J Biol Chem 272:25157–25161

Merkel LA, Rivera LM, Colussi DJ, Perrone MH (1991) Protein kinase C and vascular smooth muscle contractility: effects of inhibitors and downregulation. J Pharmacol Exp Ther 257:134–140

Mezgueldi M, Mendre C, Calas B, Kassab R, Fattoum A (1995) Characterization of the regulatory domain of gizzard calponin-interactions of the 145–163 region with F-actin, calcium-binding proteins, and tropomyosin. J Biol Chem 270:8867–8876

Mii S, Khalil RA, Morgan KG, Ware JA, Kent KC (1996) Mitogen-activated protein kinase and proliferation of human vascular smooth muscle cells. Am J Physiol Heart Circ Physiol 270:H142–H50

Mino T, Yuasa U, Nakamura F, Naka M, Tanaka T (1998) Two distinct actin binding sites of smooth muscle calponin. Eur J Biochem 251:262–268

Miriel V, Mauban J, Blaustein MP, Wier WG (1999) Local and cellular Ca^{2+} transients in smooth muscle of pressurized rat resistance arteries during myogenic and agonist stimulation. J Physiol 518.3:815–824

Missiaen L, Callewaert G, De Smedt H, Parys JB (2001) 2-Aminoethoxydiphenyl borate affects the inositol 1,4,5-trisphosphate receptor, the intracellular Ca^{2+} pump and the nonspecific Ca^{2+} leak from the non-mitochondrial Ca^{2+} stores in permeabilized A7r5 cells. Cell Calcium 29:111–116

Mita M, Walsh MP (1997) Alpha$_1$-adrenoceptor-mediated phosphorylation of myosin in rat-tail arterial smooth muscle. Biochem J 327(Pt 3):669–674

Miyazaki K, Yano T, Schmidt DJ, Tokui T, Shibata M, Lifshitz LM, Kimura S, Tuft RA, Ikebe M (2002) Rho-dependent agonist-induced spatio-temporal change in myosin phosphorylation in smooth muscle cells. J Biol Chem 277:725–734

Moreland S, Moreland RS, Singer HA (1987) Apparent dissociation between myosin light chain phosphorylation and maximal velocity of shortening in KCl depolarized swine carotid artery; effect of temperature and KCl concentration. Pflug Archiv 408:139–145

Moreland S, Antes LM, McMullen DM, Sleph PG, Grover GJ (1990) Myosin light-chain phosphorylation and vascular resistance in canine anterior tibial arteries in situ. Pflug Arch 417:180–184

Moreland SJ, Nishimura J, Van Breemen C, Ahn HY, Moreland RS (1992) Transient myosin phosphorylation at constant Ca^{2+} during agonist activation of permeabilized arteries. Am J Physiol 263:C540–C544

Morgan KG, Gangopadhyay SS (2001) Invited review: crossbridge regulation by thin filament-associated proteins. J Appl Physiol 91:953–962

Morgan JP, Morgan KG (1982) Vascular smooth muscle: the first recorded Ca^{2+} transients. Pflug Arch 395:75–77

Morgan JP, Morgan KG (1984) Stimulus-specific patterns of intracellular calcium levels in ferret portal vein smooth muscle. J Physiol 351:155–167

Morishige K, Shimokawa H, Eto Y, Hoshijima M, Kaibuchi K, Takeshita A (2001) In vivo gene transfer of dominant-negative rho-kinase induces regression of coronary arteriosclerosis in pigs. Ann N Y Acad Sci 947:407–411

Morrison DL, Sanghera JS, Stewart J, Sutherland C, Walsh MP, Pelech SL (1996) Phosphorylation and activation of smooth muscle myosin light chain kinase by MAP kinase and cyclin-dependent kinase-1. Biochem Cell Biol 74:549–557

Mukai Y, Shimokawa H, Matoba T, Kandabashi T, Satoh S, Hiroki J, Kaibuchi K, Takeshita A (2001) Involvement of Rho-kinase in hypertensive vascular disease: a novel therapeutic target in hypertension. FASEB J 15:1062–1064

Mulvany MJ, Aalkjaer C (1990) Structure and function of small arteries. Physiol Rev 70:921–961

Murphy RA (1994) What is special about smooth muscle? The significance of covalent crossbridge regulation. FASEB J 8:311–318

Nagumo H, Sasaki Y, Ono Y, Okamoto H, Seto M, Takuwa Y (2000) Rho kinase inhibitor HA-1077 prevents Rho-mediated myosin phosphatase inhibition in smooth muscle cells. Am J Physiol Cell Physiol 278:C57–C65

Naito Y, Watanabe Y, Yokokura H, Sugita R, Nishio M, Hidaka H (1997) Isoform-specific activation and structural diversity of calmodulin kinase I. J Biol Chem 272:32704–32708

Nelson MT, Cheng H, Rubart M, Santana LF, Bonev AD, Knot HJ (1995) Relaxation of arterial smooth muscle by calcium sparks. Science 270:633–637

Nelson MT, Conway MA, Knot HJ, Brayden JE (1997) Chloride channel blockers inhibit myogenic tone in rat cerebral arteries. J Physiol 502:259–264

Ngai PK, Walsh MP (1984) Inhibition of smooth muscle actin-activated myosin Mg^{2+}-ATPase activity by caldesmon. J Biol Chem 259:13656–13659

Nguyen DHD, Catling AD, Webb DJ, Sankovic M, Walker LA, Somlyo AV, Weber MJ, Gonias SL (1999) Myosin light chain kinase functions downstream of Ras/ERK to promote migration of urokinase-type plasminogen-activator cells in an integrin-selective manner. J Cell Biol 146:149–164

Nilsson H, Goldstein M, Nilsson O (1986) Adrenergic innervation and neurogenic response in large and small arteries and veins from the rat. Acta Physiol Scand 126:121–133

Nishimura J, Kolber M, Van Breemen C (1988) Norepinephrine and GTP-γ-S increase myofilament Ca^{2+} sensitivity in α-toxin permeabilized arterial smooth muscle. Biochem Biophys Res Commun 157:677–683

Nobe K, Paul RJ (2001) Distinct pathways of Ca^{2+} sensitization in porcine coronary artery: effects of rho-related kinase and protein kinase c inhibition on force and intracellular Ca^{2+}. Circ Res 88:1283–1290

Notarianni G, Gusev N, Lafitte D, Hill TJ, Cooper HS, Derrick PJ, Marston SB (2000) A novel Ca^{2+} binding protein associated with caldesmon in Ca^{2+}-regulated smooth muscle thin filaments: evidence for a structurally altered form of calmodulin. J Muscle Res Cell Motil 21:537–549

Ohanian V, Ohanian J, Shaw L, Scarth S, Parker PJ, Heagerty AM (1996) Identification of protein kinase C isoforms in rat mesenteric small arteries and their possible role in agonist-induced contraction. Circ Res 78:806–812

Omote M, Kajimoto N, Mizusawa H (1993) The ionic mechanism of phenylephrine-induced rhythmic contractions in rabbit mesenteric arteries treated with ryanodine. Acta Physiol Scand 147:9–13

Papageorgiou P, Morgan KG (1991) Intracellular free Ca^{2+} is elevated in hypertrophic aortic muscle from hypertensive rats. Am J Physiol 260:H507–H515

Parthimos D, Edwards DH, Griffith TM (1999) Minimal model of arterial chaos generated by coupled intracellular and membrane Ca^{2+} oscillations. Am J Physiol 277:H1119–H1144

Pearson RB, Wettenhall REH, Means AR, Hartshorne DJ, Kemp BE (1988) Autoregulation of enzymes by pseudosubstrate prototopes: myosin light chain kinase. Science 241:970–973

Peng H, Matchkov V, Ivarsen A, Aalkjaer C, Nilsson H (2001) Hypothesis for the initiation of vasomotion. Circ Res 88:810–815

Persechini A, Cronk B (1999) The relationship between the free concentrations of Ca^{2+} and Ca^{2+}-calmodulin in intact cells. J Biol Chem 274:6827–6830

Pucovsky V, Gordienko DV, Bolton, TB (2002) Effect of nitric oxide donors and noradrenaline on Ca^{2+} release sites and global intracellular Ca^{2+} in myocytes from guinea-pig small mesenteric arteries. J Physiol 539:25–39

Quadroni M, James P, Carafoli E (1994) Isolation of phosphorylated calmodulin from rat liver and identification of the in vitro phosphorylation sites. J Biol Chem 269:16116–16122

Rasmussen H, Takuwa Y, Park S (1987) Protein kinase C in the regulation of smooth muscle contraction. FASEB J 1:177–185

Roberts RE (2001) Role of the extracellular signal-regulated kinase (ERK) signal transduction cascade in α_2-adrenoceptor-mediated vasoconstriction in porcine palmar lateral vein. Br J Pharmacol 133:859–866

Rokolya A, Singer HA (2000) Inhibition of CaM kinase II activation and force maintenance by KN-93 in arterial smooth muscle. Am J Physiol Cell Physiol 278:C537–C545

Ruegg JC, Meisheri K, Pfitzer G, Zeugner C (1983) Skinned coronary smooth muscle: calmodulin, calcium antagonists, and cAMP influence contractility. Basic Res Cardiol 78:462–471

Ruegg JC, Pfitzer G, Zimmer M, Hofmann F (1984) The calmodulin fraction responsible for contraction in an intestinal smooth muscle. FEBS Lett 170:383–386

Ruehlmann DO, Lee C-H, Poburko D, Van Breemen C (2000) Asynchronous Ca^{2+} waves in intact venous smooth muscle. Circ Res 86:e72–e79

Sacks DB, McDonald JM (1989) Calmodulin as substrate for insulin-receptor kinase. Phosphorylation by receptors from rat skeletal muscle. Diabetes 38:84–90

Sakurada S, Okamoto H, Takuwa N, Sugimoto N, Takuwa Y (2001) Rho activation in excitatory agonist-stimulated vascular smooth muscle. Am J Physiol Cell Physiol 281:C571–C578

Sanders KM (2001) Signal transduction in smooth muscle. Invited review: mechanisms of calcium handling in smooth muscles. J Appl Physiol 91:1438–1449

Sato K, Dohi Y, Suzuki S, Miyagawa K, Takase H, Kojima M, Van Breemen C (2001) Role of Ca^{2+}-sensitive protein kinase C in PE enhancement of Ca^{2+} sensitivity in rat tail artery. J Cardiovasc Pharmacol 38:347–355

Schulman H, Hanson PI (1993) Multifunctional Ca^{2+}/calmodulin-dependent protein kinase. Neurochem Res 18:65–77

Schwartz A (1994) Molecular studies of the calcium antagonist binding site on calcium channels. Am J Cardiol 73:12B–14B

Sell M, Boldt W, Markwardt F (2002) Desynchronizing effect of the endothelium on intracellular Ca^{2+} concentration dynamics in vascular smooth muscle cells of rat mesenteric arteries. Cell Calcium 32:105–120

Sellers JR (1999) Unphosphorylated crossbridges and latch: smooth muscle regulation revisited. J Muscle Res Cell Motil 20:347–349

Senba S, Eto M, Yazawa M (1999) Identification of trimeric myosin phosphatase (PP1 M) as a target for a novel PKC-potentiated protein phosphatase-1 inhibitory protein (CPI17) in porcine aorta smooth muscle. J Biochem 125:354–362

Shaul PW, Anderson RG (1998) Role of plasmalemmal caveolae in signal transduction. Am J Physiol 275:L843–L851

Shin HM, Hyun-Dong J, Gallant C, Tao TC, Hartshorne DJ, Ito M, Morgan KG (2002) Differential association and localization of myosin phosphatase subunits during agonist-induced signal transduction in smooth muscle. Circ Res 90:546–553

Shirinsky VP, Biryukov KG, Hettasch JM, Sellers JR (1992) Inhibition of the relative movement of actin and myosin by caldesmon and calponin. J Biol Chem 267:15886–15892

Shmigol AV, Eisner D, Wray S (2001) Simultaneous measurements of changes in SR and cytosolic $[Ca^{2+}]$ in rat uterines smooth muscle cells. J Physiol 531:707–713

Siegman MJ, Butler TM, Mooers SU, Michalek A (1984) Ca^{2+} can affect V_{max} without changes in myosin light chain phosphorylation in smooth muscle. Pflugers Arch 401:385–390

Singer HA (1990) Phorbol ester-induced stress and myosin light chain phosphorylation in swine carotid medial smooth muscle. J Pharmacol Exp Ther 252:1068–1074

Singer HA, Abraham ST, Schworer CM (1996) Calcium/calmodulin-dependent protein kinase II. In: Bárány M (ed) Biochemistry of smooth muscle contraction. Academic Press, San Diego, pp 143–153

Singer HA, Benscoter HA, Schworer CM (1997) Novel Ca^{2+}/calmodulin-dependent protein kinase II γ-subunit variants expressed in vascular smooth muscle, brain, and cardiomyocytes. J Biol Chem 272:9393–9400

Sobieszek A (1977) Vertebrate smooth muscle myosin: Enzymatic and structural properties. The biochemistry of smooth muscle. Winnipeg Symposium, August 1975, pp 413–443

Soderling TR, Chang B, Brickey D (2001) Cellular signaling through multifunctional Ca^{2+}/calmodulin-dependent protein kinase II. J Biol Chem 276:3719–3722

Sohn UD, Cao W, Tang DC, Stull JT, Haeberle JR, Wang CL, Harnett KM, Behar J, Biancani P (2001) Myosin light chain kinase- and PKC-dependent contraction of LES and esophageal smooth muscle. Am J Physiol Gastrointest Liver Physiol 281:G467–G478

Somlyo AV, Franzini-Armstrong C (1985) New views of smooth muscle structure using freezing, deep-etching and rotary shadowing. Experientia 41:841–856

Somlyo AP, Somlyo AV (1998) From pharmacomechanical coupling to G-proteins and myosin phosphatase. Acta Physiol Scand 164:437–448

Somlyo AP, Somlyo AV (2000) Signal transduction by G-proteins, Rho-kinase and protein phosphatase to smooth muscle and nonmuscle myosin II. J Physiol 522.2:177–185

Somlyo AV, Goldman YE, Fujimori T, Bond M, Trentham DR, Somlyo AP (1988) Crossbridge kinetics, cooperativity, and negatively strained crossbridges in vertebrate smooth muscle. A laser-flash photolysis study. J Gen Physiol 91:165–192

Srinivasan M, Edman CF, Schulman H (1994) Alternative splicing introduces a nuclear localization signal that targets multifunctional CaM kinase to the nucleus. J Cell Biol 126:839–852

Stepien O, Marche P (2000) Amlodipine inhibits thapsigargin-sensitive $Ca(^{2+})$ stores in thrombin-stimulated vascular smooth muscle cells. Am J Physiol Heart Circ Physiol 279:H1220–H1227

Suematsu E, Resnick M, Morgan KG (1991) Change of Ca^{2+} requirement for myosin phosphorylation by prostaglandin F_{2a}. Am J Physiol Cell Physiol 261:C253–C258

Suenaga H, Kamata K (2000) Alpha-adrenoceptor agonists produce Ca^{2+} oscillations in isolated rat aorta: role of protein kinase C. J Smooth Muscle Res 36:205–218

Sutherland C, Walsh, MP (1989) Phosphorylation of caldesmon prevents its interaction with smooth muscle myosin. J Biol Chem 264:578–583

Taggart MJ (2001) Smooth muscle excitation-contraction coupling: a role for caveolae and caveolins? News Physiol Sci 16:61–65

Taggart MJ, Lee Y-H, Morgan KG (1999) Cellular redistribution of PKCα, rhoA, and ROKα following smooth muscle agonist stimulation. Exp Cell Res 251:92–101

Taggart MJ, Leavis P, Feron O, Morgan KG (2000) Inhibition of PKCα and *Rho*A translocation in differentiated smooth muscle by a caveolin scaffolding domain peptide. Exp Cell Res 258:72–81

Takahashi K, Hiwada K, Kobuku T (1988) Vascular smooth muscle calponin: a novel troponin T-like protein. Hypertension 11:620–626

Takeuchi Y, Yamamoto H, Matsumoto K, Kimura T, Katsuragi S, Miyakawa T, Miyamoto E (1999) Nuclear localization of the δ subunit of Ca^{2+}/calmodulin-dependent protein kinase II in rat cerebellar granule cells. J Neurochem 72:815–825

Tansey MG, Hori M, Karaki H, Kamm KE, Stull JT (1990) Okadaic acid uncouples myosin light chain phosphorylation and tension in smooth muscle. FEBS Lett 270:219–221

Tribe RM, Borin ML, Blaustein MP (1994) Functionally and spatially distinct Ca^{2+} stores are revealed in cultured vascular smooth muscle cells. Proc Natl Acad Sci USA 91:5908–5912

Tseng S, Kim R, Kim T, Morgan KG, Hai C-M (1997) F-actin disruption attenuates agonist-induced $[Ca^{2+}]$, myosin phosphorylation and force in smooth muscle. Am J Physiol 41:C1960–C1967

Uehata M, Ishizaki T, Satoh H, Ono T, Kawahara T, Morishita T, Tamakawa H, Yamagami K, Inui J, Maekawa M, Narumiya S (1997) Calcium sensitization of smooth muscle mediated by a Rho-associated protein kinase in hypertension. Nature 389:990–994

Van Bavel E, Mulvany MJ (1994) Role of wall tension in the vasoconstrictor response of cannulated rat mesenteric small arteries. J Physiol 477:103–115

Van Bavel E, Wesselman JPM, Spaan JAE (1998) Myogenic activation and calcium sensitivity of cannulated rat mesenteric small arteries. Circ Res 82:210–220

Van Breeman C, Chen Q, Laher I (1995) Superficial buffer barrier function of smooth muscle SR. Trends Pharmacol Sci 16:98–105

Van Zweiten PA, Pfaffendorf M (1993) Pharmacology of the dihydropyridine calcium antagonists: relationship between lipophilicity and pharmacodynamic responses. J Hypertens 11 [Suppl 6]:S3–S8

Vyas TB, Mooers SU, Narayan SR, Witherell JC, Siegman MJ, Butler TM (1992) Cooperative activation of myosin by light chain phosphorylation in permeabilized smooth muscle. Am J Physiol 263:C210–C219

Wagner PD, George JN (1986) Phosphorylation of thymus myosin increases its apparent affinity for actin but not its maximum ATPase rate. Biochemistry 25:913–918

Wang Z, Jiang H, Yang Z-Q, Chacko S (1997) Both N-terminal myosin-binding and C-terminal actin binding sites on smooth muscle caldesmon are required for caldesmon-mediated inhibition of actin filament velocity. Proc Natl Acad Sci USA 94:11899–11904

Watts SW (1996) Serotonin activates the mitogen-activated protein kinase pathway in vascular smooth muscle: use of the mitogen-activated protein kinase kinase inhibitor PD098059. J Pharmacol Exp Ther 279:1541–1550

Weber LP, Van Lierop JE, Walsh MP (1999) Ca^{2+}-independent phosphorylation of myosin in rat caudal artery and chicken gizzard myofilaments. J Physiol 516(Pt 3):805–824

Whitmarsh AJ, Davis RJ (1996) Transcription factor AP-1 regulation by mitogen-activated protein kinase signal transduction pathways. J Mol Med 74:589–607

Wier WG, Balke CW, Michael JA, Mauban JRH (2000) A custom confocal and two-photon digital laser scanning microscope Am J Physiol 278:H2150–H2156

Wills FL, McCubbin WD, Kay CM (1993) Characterization of the smooth muscle calponin and calmodulin complex. Biochemistry 32:2321–2328

Wilson DP, Sutherland C, Walsh MP (2002) Ca^{2+} activation of smooth muscle contraction. J Biol Chem 277:2186–2192

Winder SJ, Walsh MP (1990) Smooth muscle calponin: inhibition of actomyosin MgATPase and regulation by phosphorylation. J Biol Chem 265:10148–10155

Woodsome TP, Eto M, Everett A, Brautigan DL, Kitazawa T (2001) Expression of CPI-17 and myosin phosphatase correlates with Ca^{2+} sensitivity of protein kinase C-induced contraction in rabbit smooth muscle. J Physiol 535(Pt 2):553–564

Wuytack F, Raeymaekers L, De Smedt H, Eggermont JA, Missiaen L, Van Den BL, De Jaegere S, Verboomen H, Plessers L, Casteels R (1992) $Ca^{(2+)}$-transport ATPases and their regulation in muscle and brain. Ann N Y Acad Sci 671:82–91

Xiao D, Zhang L (2002) ERK MAP kinases regulate smooth muscle contraction in ovine uterine artery: effect of pregnancy. Am J Physiol Heart Circ Physiol 282:H292–H300

Xu Q, Liu Y, Gorospe M, Udelsman R, Holbrook NJ (1996) Acute hypertension activates mitogen-activated protein kinases in arterial wall. J Clin Invest 97:508–514

Yamboliev IA, Hedges JC, Mutnick JL-M, Adam LP, Gerthoffer WT (2000) Evidence for modulation of smooth muscle force by the p38 MAP kinase/HSP27 pathway. Am J Physiol Heart Circ Physiol 278:H1899–H1907

Yoshikawa H, Taniguchi S-I, Yamamura H, Mori S, Sugimoto M, Miyado K, Nakamura K, Nakao K, Katsuki M, Shibata N, Takahashi K (1998) Mice lacking smooth muscle calponin display increased bone formation that is associated with enhancement of bone morphogenetic protein responses. Genes Cells 3:685–695

Zang W-J, Balke CW, Wier WG (2001) Graded α_1-adrenoceptor activation of arteries involves recruitment of smooth muscle cells to produce "all or none" Ca^{2+} signals. Cell Calcium 29:327–334

Zelcer E, Sperelakis, N (1982) Spontaneous electrical activity in pressurized small mesenteric arteries. Blood Vessels 19:301–310

Zhang J, Wier WG, Blaustein MP (2002) Mg^{2+} blocks myogenic tone but not K^+-induced constriction: role for SOCs in small arteries. Am J Physiol 283:H2692–H2705

Zhong H, Minneman KP (1999) α_1-Adrenoceptor subtypes. Eur J Pharm 375:261–276

Zhuang S, Mabuchi K, Wang C-LA (1996) Heat treatment could affect the biochemical properties of caldesmon. J Biol Chem 271:30242–30248

Zimmermann B, Somlyo AV, Ellis-Davis GCR, Kaplan JH, Somlyo AP (1995) Kinetics of prephosphorylation reactions and myosin light chain phosphorylation in smooth muscle. J Biol Chem 270:23966–23974

Zou H, Ratz PH, Hill MA (2000) Temporal aspects of Ca^{2+} and myosin phosphorylation during myogenic and norepinephrine-induced arteriolar constriction. J Vasc Res 37:556–567

Rev Physiol Biochem Pharmacol (2003) 150:140–160
DOI 10.1007/s10254-003-0020-2

G. Ahnert-Hilger · M. Höltje · I. Pahner · S. Winter · I. Brunk

Regulation of vesicular neurotransmitter transporters

Published online: 27 September 2003
© Springer-Verlag 2003

Abstract Neurotransmitters are key molecules of neurotransmission. They are concentrated first in the cytosol and then in small synaptic vesicles of presynaptic terminals by the activity of specific neurotransmitter transporters of the plasma and the vesicular membrane, respectively. It has been shown that postsynaptic responses to single neurotransmitter packets vary over a wide range, which may be due to a regulation of vesicular neurotransmitter filling. Vesicular filling depends on the availability of transmitter molecules in the cytoplasm and the active transport into secretory vesicles relying on a proton gradient. In addition, it is modulated by vesicle-associated heterotrimeric G proteins, $G\alpha o2$ and $G\alpha q$, which regulate VMAT activities in brain and platelets, respectively, and may also be involved in the regulation of VGLUTs. It appears that the vesicular content activates the G protein, suggesting a signal transduction form the luminal site which might be mediated by a vesicular G-protein coupled receptor or, as an alternative, possibly by the transporter itself. These novel functions of G proteins in the control of transmitter storage may link regulation of the vesicular content to intracellular signal cascades.

Introduction

Neurotransmitters are key molecules of neurotransmission. They are concentrated first in the cytosol and then in small synaptic vesicles of presynaptic terminals by the activity of specific neurotransmitter transporters of the plasma and the vesicular membrane, respectively. Following an action potential, synaptic vesicles fuse with the plasma membrane and release their transmitter content as multimolecular packets into the synaptic cleft, where the transmitter reacts with receptors at the postsynaptic site. It has been shown that postsynaptic responses to single neurotransmitter packets vary over a wide range. This might be due to variations in the postsynaptic receptor population or modulations of their

G. Ahnert-Hilger (✉) · M. Höltje · I. Pahner · S. Winter · I. Brunk
Institut für Anatomie und Neurowissenschaftliches Zentrum der Charité,
Humboldt-Universität zu Berlin, Philippstr. 12, 10115 Berlin, Germany
e-mail: gudrun.ahnert@charite.de · Tel.: +49-30-450528276 · Fax: +49-30-450528912

affinities. Alternatively, the amount of transmitter molecules vary per single vesicular quantum. Both parameters may vary and contribute to synaptic strength and synaptic plasticity. Changes in synaptic plasticity have been confined more to postsynaptic events, while the contribution of a regulated vesicular filling is probably underestimated. The present review will focus on the regulation of vesicular neurotransmitter filling which depends on the availability of transmitter molecules in the cytoplasm and the active transport into secretory vesicles relying on a proton gradient, but which is also linked to intracellular signaling pathways.

Factors influencing synaptic strength and neurotransmitter content of secretory vesicles

Communication within the nervous system and between neurons and nonneuronal tissue is mainly sustained by synaptic transmission. Synaptic transmission involves a highly regulated fusion process between synaptic vesicles and the plasma membrane initiated by the formation of the ternary SNARE complex (Sutton et al. 1998; Jahn and Südhof 1999; Lin and Scheller 2000). This fusion process is more or less common to all vesicles irrespective of the nature of the stored transmitter. The availability of the individual SNARE proteins and their interaction with presynaptic cytoskeletal elements determine the number of vesicles released and consequently add to synaptic strength. These parameters also define the readily releasable and the reserve pool of synaptic vesicles (Pieribone 1995). Short-term depression is caused when rapid action potentials expend the readily releasable pool more rapidly than it recovers (Stevens and Wesseling 1999). Depending on the type of synapse, a more or less rapid refilling of vesicles in the readily releasable pool may result in more or less pronounced short-term depression. Vesicle pool organization emerges during neuronal development. Nevertheless, the organization of vesicles in either pool does not appear to depend directly on their transmitter content (Mozhayeva et al. 2002). At least at the neuromuscular junction, some types of depression are due to a reduced refilling of vesicles with neurotransmitter (Elmquist and Quastel 1965). Even almost empty vesicles appear to have the same chance to fuse with the membrane (Travis et al. 2000; van der Kloot et al. 2000, 2002). Therefore, transmitter content and fusion capacity are regulated by different mechanisms. Irrespective of whether a vesicle is able to immediately fuse with the plasma membrane (as a member of the readily releasable pool) or is recruited from the reserve pool, the transmitter content directly influences the postsynaptic response. Thus, in addition to the availability of synaptic vesicles for fusion, synaptic strength also depends on the concentration of transmitter in the synaptic cleft which directly correlates with the filling stage of presynaptic vesicles.

Secretory vesicles and their neurotransmitter transporters

Various secretory vesicles store low molecular weight neurotransmitters. These include small synaptic vesicles (SSV) in neuronal terminals, SSV-analogues also referred to as small synaptic-like microvesicles in neuroendocrine cells, and dense core and large dense core vesicles occurring beside SSV in neurons and neuroendocrine cells. Dense core vesicles also contain a variety of peptides as cotransmitters. Neurotransmitters are concentrated in secretory vesicles by means of vesicular neurotransmitter transporters. So far, seven transmitter-specific vesicular neurotransmitter transporters have been identified.

Vesicular monoamine transporters

Two structurally related but pharmacologically distinct vesicular monoamine transporters (VMAT) are known. VMAT1 has been cloned from PC12 cells (Liu et al. 1992) and VMAT2 from rat brain (Liu et al. 1992; Erickson et al 1992; Liu et al. 1994). VMAT2 is the dominant transporter in brain, but also occurs in a variety of peripheral cells like sympathetic neurons, enterochromaffin-like cells (Peter et al. 1995; Erickson et al. 1996), and also in blood platelets (Lesch et al. 1993). By contrast, VMAT1 appears to occur only in the periphery, at least in adult individuals. During development of the rat central nervous system, a transient coexpression of VMAT1 and VMAT2 has been reported for limbic structures, basal ganglia, and some hypothalamic structures (Hansson et al. 1998), but the tissue-specific coding for one of the transporters appears to be developmentally predetermined (Schütz et al. 1998). Both transporters accept monoamines such as serotonin, dopamine, noradrenaline, and adrenaline at comparable concentrations with micromolar K_m values for VMAT1 and submicromolar K_m values for VMAT2 in rat (Peter et al. 1994) and humans (Erickson et al. 1996). VMAT2 also transports histamine, barely recognized by VMAT1, and is ten times (rat; Peter et al. 1994) to 100 times (human; Erickson et al. 1996) more sensitive to tetrabenazine. VMAT2 has a general higher affinity for monoamines than VMAT1, which may be required for rapidly recycling SSV in brain in contrast to more slowly filling secretory granules in the adrenal medulla (Peter et al. 1994).

Vesicular acetylcholine transporter

The vesicular acetylcholine transporter VAChT has been cloned first from *Caenorhabditis elegans* as UNC-17 (Alfonso et al. 1993) followed by its identification in *Torpedo* (Varoqui et al. 1994) and rat (Erickson et al. 1994). VAChT is closely related to the VMATs and has a unique relationship to the gene of choline acetyltransferase (Usdin et al. 1995). It occurs exclusively on SSV of cholinergic terminals in the peripheral and central nervous system (Weihe et al. 1996). In tegmental neurons, high concentrations of VAChT have also been found to be associated with tubulovesicular structures in dendrites, suggesting an additional dendritic release of actelycholine (Garzon and Pickel 2000). Despite its close relationship to VMATs, VAChT has an almost 1,000-fold higher millimolar range K_m value for acetylcholine transport (Varoqui and Erickson 1996).

Vesicular glutamate transporters

Three vesicular glutamate transporters, VGLUT1 (Bellocchio et al. 2000; Takamori et al. 2000a), VGLUT2 (Fremeau et al. 2001; Bai et al. 2001; Takamori et al. 2001; Hayashi et al. 2001), and VGLUT3 (Gras et al. 2002; Takamori et al. 2002; Fremeau et al. 2002; Schäfer et al. 2002) have been cloned. VGLUT1 and 2 have been originally described as brain-specific or differentiation-associated, Na^+-dependent inorganic phosphate transporters, BNPI (Ni et al. 1994) or DNPI (Hisano et al. 2000), respectively, but later have been shown to transport glutamate with high avidity and specificity. Generally, the three VGLUTs exhibit a higher K_m to their substrate than VMATs, approximately 1–2 mM for both VGLUT1 and VGLUT2 (Bellocchio et al. 2000; Gras et al. 2002) and about 0.6 mM for VGLUT3 (Gras et al. 2002). All three are, however, very specific for glutamate and do not accept other amino acids like aspartate (Takamori et al. 2000a, 2001; Gras et al.

2002). The apparent affinity of glutamate to vesicular glutamate transporters is 1–2 orders of magnitude lower than to its various plasma membrane transporters. Nevertheless, apparently a high cytoplasmic concentration of glutamate of 100 mM guarantees a sufficient loading of vesicles with the transmitter (for review, see Danbolt 2001). High amounts of extracellular glutamate, which may excessively activate glutamate receptors, are harmful for neurons and the balance of the neuronal network. Therefore, it is important to keep the extracellular concentration low, which explains the several thousand-fold gradient over the plasma membrane, especially in nerve terminals (Danbolt 2001 and references cited therein). Cytosolic glutamate, in contrast, is harmless; therefore, the concentration gradient between the cytosolic and the intravesicular compartment reaches only a factor of ten. This, in turn, is in contrast to VMATs. Their substrate affinity is high because an increased concentration of monoamines may lead to oxidized products interfering with mitochondrial enzymes. In addition to their function as vesicular glutamate transporters, VGLUT1 and VGLUT2 appear to transport phosphate into the cytoplasm of nerve terminals (Ni et al. 1994) probably when integrated in the plasma membrane. So far, it is unclear how the decision between either function is regulated.

VGLUT1 and VGLUT2 have a distinct and mutually exclusive distribution in brain, with VGLUT1 being the dominant transporter in cortex, hippocampus, and cerebellum (Fremeau et al. 2001; Fujiyama et al. 2001; Kaneko and Fujiyama 2002), and VGLUT2 in thalamic and hypothalamic regions (Hisano et al. 2000; Sakata-Haga et al. 2001; Fujiyama et al. 2001; Fremeau et al. 2002; Kaneko and Fujiyama 2002). In addition, VGLUT2 is expressed in the pineal gland and in α-cells of Langerhans islets, suggesting a role in endocrine function (Hayashi et al. 2001). A strict separation between VGLUT1- and VGLUT2-containing terminals has been shown for cerebellar cortex where parallel fiber terminals contain VGLUT1, whereas climbing fiber terminals have VGLUT2 (Fremeau et al. 2002). In contrast, VGLUT3 is found in serotonergic, cholinergic (Fremeau et al. 2002; Gras et al. 2002; Schäfer et al. 2002), and GABAergic terminals (Fremeau et al. 2002), suggesting a role for glutamate as cotransmitter in these nerve terminals.

Vesicular GABA or inhibitory amino acid transporter

The vesicular GABA transporter (VGAT), also referred to as vesicular inhibitory amino acid transporter (VIAAT), was cloned in parallel from *Caenorhabditis elegans* (McIntire et al. 1997) and from mouse (Sagné et al. 1997). VGAT also transports glycine; it exhibits a very low affinity for its substrate GABA (in the millimolar range) and an even lower for glycine (Sagne et al. 1997; McIntire et al. 1997; Bedet et al. 2000). Its presence on a subset of synaptic vesicles was confirmed by immunoisolation of GABA-specific vesicles from rat brain using a VGAT-antibody (Takamori et al. 2000b). VGAT has been found in terminals of GABAergic and gycinergic neurons, suggesting that it is the main vesicular transporter for inhibitory transmitters in brain (Chaudry et al. 1998; Dumoulin et al. 1999).

Cytoplasmic concentration of transmitters in the presynaptic terminal

Vesicular filling due to the various vesicular transmitter transporters described above depends on the availability of the respective transmitter, i.e., its concentration in the presynaptic cytosolic compartment. The synthesizing and metabolizing enzymes specific for monoamines, GABA, and glutamate sustain a certain level in the presynaptic terminal. Changes in the concentration of these enzymes or their inhibition by drugs directly influence the cytoplasmic transmitter concentration. For example, an inhibition of the GABA degradating enzyme GABA transaminase increases the GABA concentration in the presynaptic terminal (Engel et al. 2001) and tissue monoamine concentrations are increased in mice deficient for monoamine oxidase A (Cases et al. 1995).

Rapid and localized changes of neurotransmitter concentrations occur beneath the presynaptic membrane by the activity of plasma membrane transporters to clear the synaptic cleft from released neurotransmitter. Two distinct gene families encoding plasma membrane transporters have been identified. One is the sodium dependent transporters, including those for dopamine, noradrenaline or serotonin, and GABA and Glycin. Evidence emerged that these transporters undergo a rapid regulation of their translocation to the plasma membrane. The dopamine transporter DAT is rapidly cleared from the plasma membrane by endocytosis when activating PKC (Daniels and Amara 1999). Transport activities are negatively regulated by syntaxin1A as shown for the GABA transporter GAT1 (Deken et al. 2000), the glycin transporters GLYT1 and GLYT2 (Geerlings et al. 2000), and the serotonin transporter SERT (Haase et al. 2001). The second family codes for transporters of excitatory amino acids like glutamate and aspartate. A variety of neuronal and glial plasma membrane transporters (excitatory amino acid transporters, EAATs) with varying substrate specificity assure the rapid clearance of glutamate from the synaptic cleft (for review, see Danbolt 2001). The efficiency of transmitter transport into the cytosolic compartment may be regulated by sorting more or less transporter molecules to the plasma membrane. It may be that transmitter transporters reside on vesicles beneath the plasma membrane from where they can be rapidly recruited by membrane fusion and so modulate the duration of postsynaptic responses. This has recently been shown for GAT1, which is rapidly recycled upon Ca^{2+}-dependent exocytosis and stored in a distinct class of vesicles lacking synaptophysin and VGAT (Deken et al. 2003).

The activity of metabolizing and catabolizing enzymes as well as plasma membrane transporters add to the amount of transmitter molecules available in the cytosol of the terminal and indirectly to the amount of transmitter to be concentrated inside synaptic vesicles. Accordingly, mice lacking the serotonin transporter SERT have reduced serotonin concentrations in different nervous tissues and adrenal glands (Vogel et al. 2003).

Variability in the transmitter content of individual vesicles

Data on the intravesicular concentrations of neurotransmitters mostly rely on the respective miniature inhibitory or excitatory postsynaptic currents, where receptor affinity, morphology of the synaptic cleft, and distance from the receptors add to the signal analyzed and thus obscure the amount of transmitter released by a single vesicle. Amperometry, which is used for the detection of monoamines, overcomes this problem because the amount of transmitter released by a given vesicle may be measured directly. By correlating amperometric charge and vesicular size, Bruns and colleagues showed that the variations in released quanta are due to variations in the volume of the respective secretory vesicles. On average, the transmitter concentrations in SSV and LDCV are similar, being 270 mM with LDCV exhibiting a greater variability than SSV (Bruns et al. 2000).

Probably unstimulated neurons that have not experienced changes in their environment constitutively set their vesicular transmitter content according to a balanced equilibrium leading to a similar transmitter concentration in all vesicles (Bruns et al. 2000). Variations in the availability of transmitter due to changes in synthesis or metabolism may influence transmitter release if vesicles could accept more. Variations in quantal size have been documented at the frog neuromuscular junction, leading to the idea that vesicular filling may vary (van der Kloot 1991). These variations could be described either by a set point or by a steady state model, suggesting that vesicles can only accept a certain amount or get more filled if more neurotransmitter is available, respectively (Williams 1997). Vesicles with varying transmitter content have the same chance to fuse with the plasma membrane as has been reported for cholinergic vesicles (van der Kloot et al. 2000) and monoamine storing vesicles in VMAT2-deficient mice (Travis et al. 2000). Therefore, fusion competence and transmitter filling are regulated separately.

The impact of variations of the vesicular transmitter content has been investigated for the monoaminergic system in detail. Offering the dopamine precursor DOPA to PC 12 cells (Pothos et al. 1998a) or midbrain dopaminergic neurons (Pothos et al. 1998b) increases the amount of transmitter released as measured by amperometry. Similarly, an increase in the amount of transmitter transporters as shown for VAChT (Song et al. 1997), and VMAT2 (Pothos et al. 2000) increases the stimulated release of neurotransmitters. All these data suggest that secretory vesicles can accept more transmitter if conditions permit. Whether the vesicular concentration or the volume changes is not yet clear, and controversial data can be obtained from the literature. Studies on PC 12 cells suggest that increasing the amount of transmitter increases the volume of secretory granules, while decreasing the vesicular filling by reserpine shrinks the vesicles (Colliver et al. 2000). On the other hand, a 48-h incubation of leech neurons with reserpine appeared to not change the morphology of serotonin storing vesicles (Bruns et al. 2000). Accordingly, changes in the acetylcholine content did not change the size of secretory vesicles (Van der Kloot et al. 2002). Therefore, it appears that smaller vesicles are not completely filled and can change their luminal transmitter concentration. Whether special large dense core vesicle may have additional regulatives allowing them to change their volume is not yet clear.

Role of postsynaptic receptors for synaptic strength

The number of postsynaptic receptors and their affinity for the respective neurotransmitter are certainly important for synaptic strength. Regulation of vesicular filling by means other than constitutive processes would be pointless if postsynaptic receptors following an exocytotic event are generally saturated by peak concentrations of neurotransmitters in the synaptic cleft. Therefore, the question is: "Do the conditions in the synaptic cleft with respect to transmitter concentration and presence allow a saturation of all post- and presynaptic receptors?" And secondly, "Is a synapse a unit on its own or does crosstalk between neighboring synapses exist?"

The maximum fraction of receptors that bind transmitter in the likely range of peak transmitter concentration depends on the time course and peak concentration of the individual transmitter in the synaptic cleft. Mostly, the transmitter is not present long enough in the synaptic cleft—below 1 ms as predicted from kinetic models for the $GABA_A$-receptor (Frerking and Wilson 1996)—to permit equilibration of receptor binding. Receptor occupancy may differ between types of synapses and even among different sites on the same cell. Postsynaptic responses depend on the affinity of the respective receptors. Receptors with constants in the low millimolar range will probably not activate under certain conditions, but could be activated if transmitter quanta are increased (Frerking and Wilson 1996). In addition, clearance of the synaptic cleft first depends on the activity of plasma membrane transporters (in the case of acetylcholine on the activity of the acetylcholine esterase) and second on the synaptic morphology, which may allow or prevent diffusion and spillover of transmitter (Danbolt 2001). Nonsaturation of postsynaptic receptors by single presynaptic events have been described for NMDA-receptors (Mainen et al. 1999; Ishikawa et al. 2002), AMPA-receptors (Ishikawa et al. 2002), and $GABA_A$-receptors (Hajos et al. 2000).

Synaptic crosstalk may be relevant to phenomena such as long-term potentiation or depression in which the associativity and cooperativity of synapses are important. Crosstalk could be predicted by calculating glutamate diffusion in correlation to the kinetic parameters of, i.e., glutamate receptors and plasma membrane transporters, and may apply to a variety of synapses, including mossy fiber synapses in hippocampus or cerebellum (Barbour and Häusser 1997). Transmitter crosstalk preferentially activates high affinity receptors, which often produce long-lasting responses like NMDA or metabotropic receptors. It might represent an amplification mechanism when many synaptic contacts are formed by the same presynaptic fiber as in synapses between climbing fibers and Purkinje cell dendrites. By diffusion of the transmitter away from the release site, a background tuning is obtained (Barbour and Häusser 1997) as probably being the principle for a variety of monoaminergic synapses (for review, see Agnati et al. 1995).

The evidence for postsynaptic receptors being not saturated and for an existence of spillover of transmitter makes a regulation of vesicular filling feasible.

Regulation of vesicular transporter activity

Many vesicular transporters have been characterized at the molecular level; however, little is known about whether they are regulated by means of factors other than availability of the transmitter and ion gradients over the vesicular membrane. While changes in the cytosolic transmitter concentration may directly influence the vesicular filling stage, it is not clear whether the filling of a given vesicle is only constitutively regulated or whether it can be also modified in accordance with other signals from the extracellular environment. In other words, can a neuron increase or decrease the filling stage of its vesicles?

Influence of proton and ion gradients

Generally, the activity of all vesicular transporters depends on an electrochemical gradient ($\Delta\mu H^+$) over the vesicular membrane extensively described for monoamine uptake into chromaffin granules (Johnson 1988). The gradient consists of a proton gradient (ΔH^+) built up by the activity of the vacuolar H^+-ATPase, which generates an intravesicular pH of about 5.6 and an electrical gradient ($\Delta\psi$) that—due to the increase of positive charges in the vesicle lumen—results in a positive membrane potential towards cytosol. The electrochemical gradient supplies the driving force for concentrating transmitters in the vesicle lumen. These principle also apply to transmitter packaging into small synaptic vesicles, where the differences in ion gradients necessary for optimal GABA, glutamate, or dopamine uptake have been realized before the molecular identification of the respective transporters (Hell et al. 1988, 1990; Maycox et al. 1990). The various transporters differ in that they rely either more on ΔH^+ or on $\Delta\psi$, or equally on both. VMATs and VAChT activity strictly depends on (ΔH^+) with less influence of $\Delta\psi$, while VGLUTs depend more on $\Delta\psi$, and VGAT on both (Johnson 1988; Hell et al. 1990; Maycox et al. 1990; Schuldiner et al. 1995; Reimer et al. 1998). Physiologically, ΔH^+ and $\Delta\psi$ drive as $\Delta\mu H^+$ the vesicular transmitter uptake. Increasing vesicular chloride concentration by opening of vesicular chlorid channels increases ΔH^+ while decreasing $\Delta\psi$. A decrease in chloride intake increases $\Delta\psi$ and decreases ΔH^+ due to reduced proton accumulation, since under this condition the positive charges in the vesicular lumen are not neutralized by chloride ions (Maycox et al. 1990; Reimer et al. 2001). VMATs and VAChT transport one transmitter molecule in exchange for two protons; VGLUTs and VGAT exchange one proton for one transmitter molecule (Schuldiner et al. 1995; Reimers et al. 2001). In addition, glutamate acidifies the vesicular lumen comparable to chloride since the negatively charged glutamate ions result in an increased uptake of protons via ATPase. To balance charge, an additional chloride is probably transported from the vesicular lumen to the cytosol (Maycox et al. 1990; Wolosker et al. 1996). A large $\Delta\psi$ apparently increases the affinity of VGLUTs towards their substrate, but a large ΔH^+ positively influences the trapping of glutamate inside the vesicle (Wolosker et al. 1996). At a given number of transporter molecules, the transmitter concentration inside the vesicle thus depends on the availability of the transmitter and—depending on the respective transporter—on ΔH^+ and $\Delta\psi$. The highest gradient for transmitters are sustained by VMATs which concentrate monoamines by a factor of 10^4 or 10^5. In resting LDCV and SSV of the leech synapse, an average serotonin concentration of 270 mM was measured (Bruns et al. 2000). In chromaffin granules additional intravesicular components, i.e., the chromogranins, add to an effective trapping, yielding catecholamine concentration over 1 M (Johnson 1988; Schuldiner et al. 1995) The other transporters like VGLUTs and VGAT concentrate the transmitters only tenfold or less over the cytoplasmic concentration yielding an intravesicular concentration of about at least 100 mM for glutamate (Danbolt 2001).

Vesicle-associated G proteins as modulators of vesicular neurotransmitter content

Heterotrimeric G proteins consisting of an α-, a β-, and a γ-subunit are molecular switches coupling seven membrane spanning receptors to intracellular effector systems. So far, 23 α-, 5 β-, and 11 γ-subunits have been described which give more than 1,000 possible combinations conferring specificity to receptor G protein and G protein-effector interactions. Upon receptor activation, the Gα-subunit exchanges a GDP against a GTP and dissociates from its $\beta\gamma$-subunit. Both activated Gα- as well as G$\beta\gamma$-subunits interfere with various intracellular signal cascades. The Gβ-subunits are considered to be essential for the functional coupling of Gα-subunits with receptors. Gβ5 appears to be an exception, as it interacts with G-protein effectors in the absence of a γ-subunit (Sondek and Siderovsky 2001). So far, only Gαo (Jiang et al. 1998) and Gαq (Offermanns 1999) have been associated with central nervous system defects in genetic ablation studies. In this respect, it is remarkable that only scarce information about upstream and downstream signals of Gαo exists although it comprises about 1% of total protein in brain (Jiang et al. 1998; Dhingra et al. 2002).

Apart from their localization at the plasma membrane, G proteins are also found on various endomembrane systems, playing a role in membrane traffic and functional properties of intracellular organelles. While many data exist on the signal transduction from plasma membrane receptors to the respective G proteins and their addressed signal cascades, the activation of organelle-associated G proteins is less well understood. Three ways of activation may be possible. First, association of G protein subunits with endomembranes may be an intermediate step during activation of plasma membrane receptors. Second, organelle-associated G proteins could be activated by cytosolic factors. As a third possibility, G proteins reside as functional heterotrimers on endomembranes like secretory vesicles, where they also might well be activated from the luminal side of organelles (Nürnberg and Ahnert-Hilger 1996) and control vesicular properties like the filling stage (see "Vesicular filling as a regulator"). As a prerequisite for the last case, G-protein heterotrimers should be present on all types of transmitter-storing vesicles, and subtypes of secretory vesicles might differ in their subunit composition.

G proteins of different subunit composition on vesicular subpopulations

An association of G-protein subunits with secretory vesicles was first seen for Giα-subunits in synaptic terminals of basal ganglia (Aronin and DiFiglia 1992). A vesicle subtype specific association of various Gαi and Gαo-subunits was then obtained when analyzing purified secretory vesicles from bovine adrenal medulla and rat brain (Ahnert-Hilger et al. 1994). Subclasses of SSV differ in their neurotransmitter transporter, which may also be reflected by different subsets of G proteins. During subcelluar fractionation or immunoisolation, G-protein subunits may dissociate, obscuring a specific and differential association with subtypes of secretory vesicles. Quantitative postembedding immunogold electron microscopy using antibodies against the various vesicular transmitter transporters overcomes these problems. On chromaffin granules of the rat adrenal medulla Gαo2, Gαq, and Gβ-subunits, mostly Gβ2, could be identified and colocalized with either VMAT1 (see Fig. 1) or VMAT2 (Pahner et al. 2002). Since the Gβ-subunits 1–4 have to be tightly bound to a Gγ-subunit, these data indicate that chromaffin granules are equipped with functional G-protein heterotrimers which regulate VMAT activity (Pahner et al. 2002). In addition, Gαo2 is also seen on VMAT2 containing SSV in serotonergic terminals (Höltje et al. 2000). When quantifying G-protein subunits on SSV in different synaptic terminals defined by the expression of either VGLUTs or VGAT, it turned out that even glutamatergic vesicles differ in their G-protein subunit profile depending on the VGLUT subtype. Fig. 2 shows the double immunogold labeling of cerebellar parallel fiber terminals and Schaffer collateral terminals identified by the presence of VGLUT1 and cerebellar climbing fiber terminals identified by the presence of VGLUT2 with antibodies recognizing various Gβ-subunits (Gβ1–4) or specifically Gβ2 (Pahner et al. 2003). Figure 3 compares the amounts of vesicle-associated Gαo2 and Gβ-subunits in parallel fiber (VGLUT1) and climbing fiber (VGLUT2) terminals. In both types of glutamatergic terminals, equal amounts of Gαo2, Gβ1–4, and Gγ7 are found to be associated with SSV. However, only SSV in parallel fiber terminals have Gβ2. SSV of GABAergic basket cell terminals exhibit less Gαo2 compared to the two glutamatergic terminals and also less Gβ2 compared to parallel fiber terminals (Pahner et al. 2003).

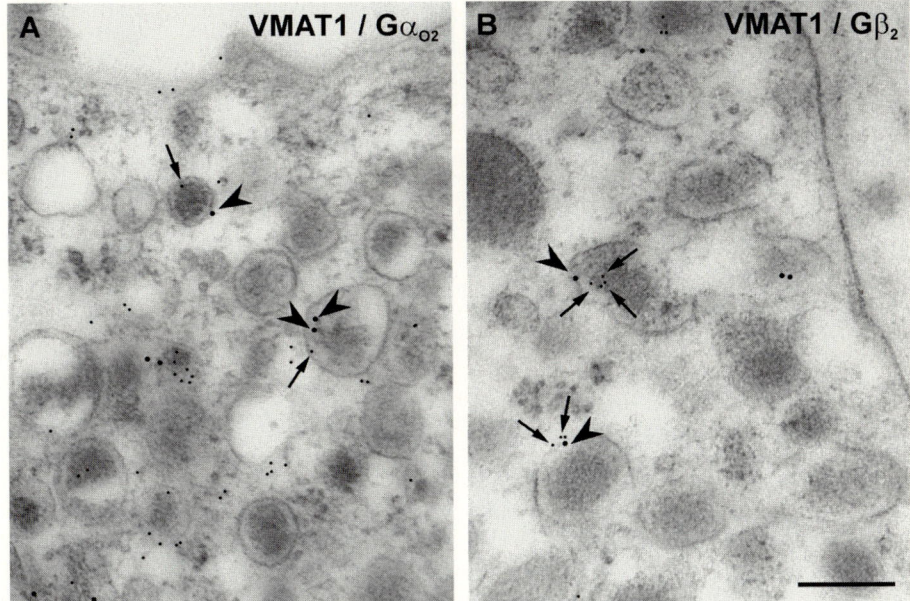

Fig. 1A, B G-protein subunits on chromaffin granules of the rat adrenal medulla. Double immunogold labeling for VMAT1 (large gold particles, several indicated by *arrowheads*) and either Gαo2 (**A**) or Gβ2 (**B**; small gold particles, several indicated by *arrows*) was performed as given (Pahner et al. 2002). VMAT1 and G-protein subunits are found to be colocalized on granules. *Scale bar*, 200 nm

These data provide strong evidence for functional G-protein heterotrimers on SSV. Furthermore, they indicate that different subsets of SSV are characterized by specific G-protein subunit combinations depending on the respective vesicular transmitter transporter.

Gα-subunits as regulators of VMAT and VGLUT activity

G-protein heterotrimers reside on secretory vesicles but so far only Gα-subunits appear to affect vesicular transmitter transporter activity. In permeabilized PC 12 cells, VMAT1 activity is downregulated by nonhydrolysable GTP-analogs, GTPγS, and GMppNp, which activate G-protein α-subunits, but not by GDPβS, which stabilizes the inactivated stage. The VMAT1 downregulating effects of GTP-analogues can be prevented by preincubating the cells with pertussis toxin. Downregulation is also observed when using activated Gαo2. Neither activated Gαi1 and Gαi2 nor Gαo1 sustain the inhibition of VMAT1 activity in PC 12 (Ahnert-Hilger et al.1998) and human BON cells (Höltje et al. 2000). Similarly, VMAT2 activity is exclusively downregulated by Gαo2, irrespective of whether the transporter resides on large dense core or small synaptic vesicles. Finally, using SSV preparations from rat or mouse brain, VMAT2 activity can be inhibited by activated Gαo2, but not when using Gαo1 (Höltje et al. 2000, see Fig. 4). The same applies to VMAT1 when using a chromaffin granule preparation from the rat adrenal medulla (Pahner et al. 2002). Taken together, vesicle-associated Gαo2 regulates VMAT activity.

Blood platelets may be regarded as a very reduced model for serotonergic neurons. They are the major storage sites of serotonin apart from brain. Platelets take up serotonin

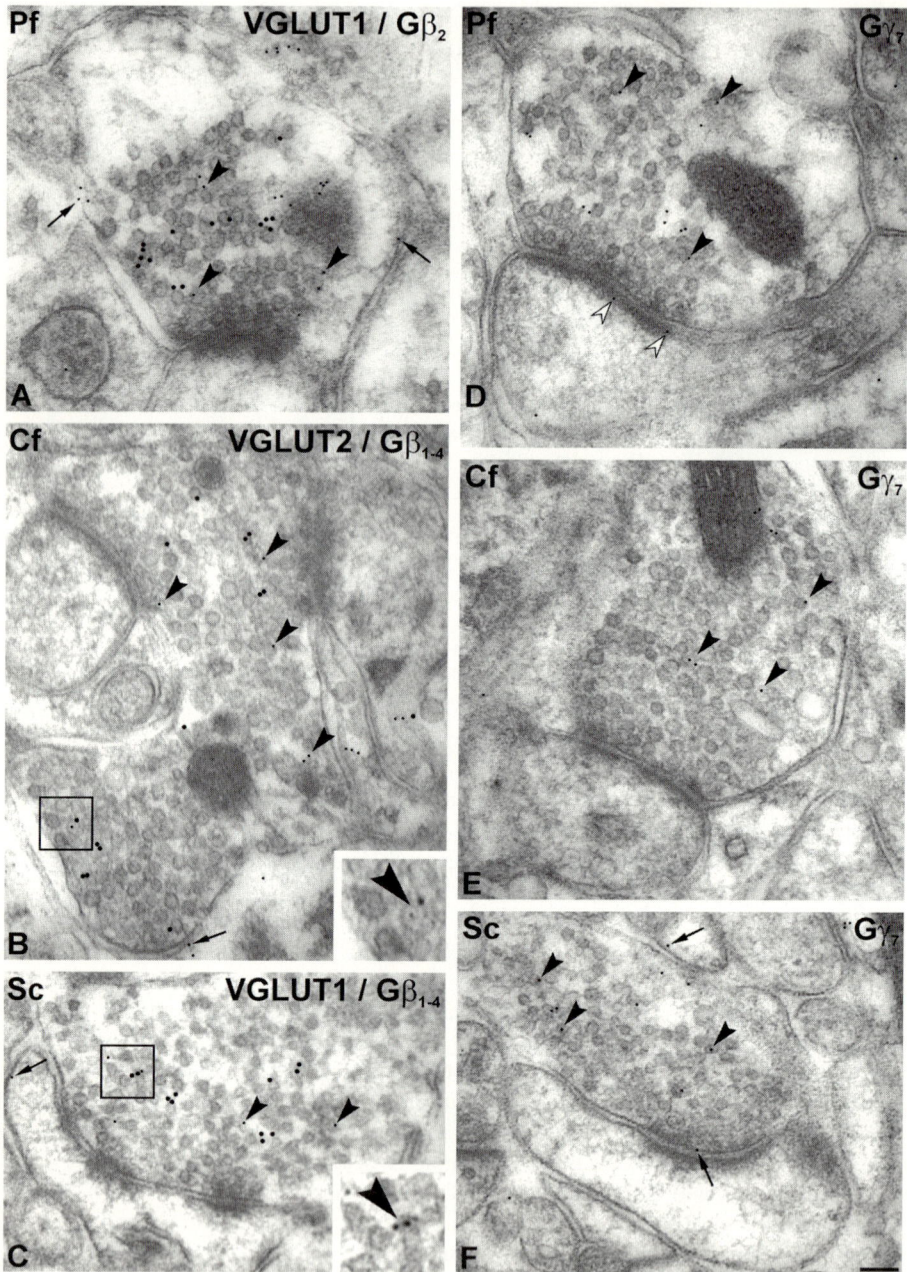

Fig. 2A–F G-protein subunits in different types of glutamatergic terminals. Three types of glutamatergic terminals characterized either by VGLUT1 (cerebellar parallel fiber terminals, Pf and hippocampal Schaffer collateral terminals, *Sc*) or VGLUT2 (cerebellar climbing fiber terminals, *Cf*) contain Gβ- or Gγ-subunits on their small synaptic vesicles. Double immunogold labeling (**A–C**) indicates that the transporter (large gold particles) and the Gβ-subunits (small gold particle, several indicated by *arrowheads*) occasionally reside on the same vesicle (*arrowheads* in the *insets* in **B** and **C**). Gγ7 is associated with small synaptic vesicles in all three types of glutamatergic terminals (**D–F**). Immunogold signals for G-protein subunits associated with the plasma membrane or a postsynaptic density are indicated by *arrows* (**A–C, F**) or *open arrowheads* (**D**), respectively. *Scale bar*, 50 nm. (From Pahner et al. 2003, with permission)

Fig. 3A–C Direct comparison of Gα-, Gβ-, and Gγ-subunits on SSV in two glutamatergic terminal types. Association of G-protein subunits was analyzed in parallel fiber (*Pf*) or climbing fiber terminals (*Cf*). *Columns* represent fractions of immunolabeled vesicles (percentage of total number of vesicles) and combine data obtained from two animals. The *star* denotes statistical significance (*p*<0.05) according to the exact version of Pearson's χ^2 test. (Modified from Pahner et al. 2003, with permission)

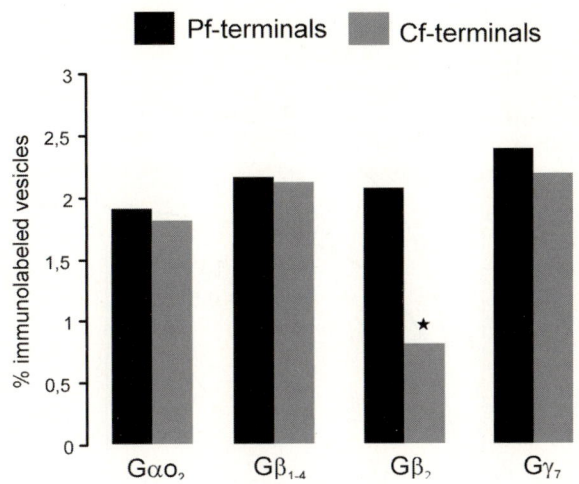

Fig. 4 Gαo2 but not Gαo1 downregulates serotonin uptake into a synaptic vesicle preparation from rat brain. Crude synaptic vesicles (*LP2*) were loaded with [³H] serotonin in the presence of either 100 µM GMpp-Np,10 nM Gαo2, or 20 nM Gαo1 for 10 min. Values are expressed as percent of reserpine-sensitive uptake performed in the absence of GMppNp and set as 100%. Data from three independent experiments (mean±SD) are combined. (Modified from Höltje et al. 2000, with permission)

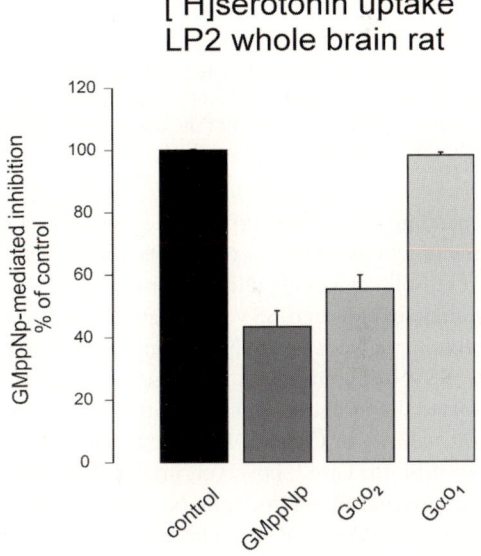

by the plasma membrane transporter SERT and concentrate it by VMAT2 activity in secretory vesicles, i.e., large dense bodies. However, platelets do not contain Gαo2, which is mainly expressed in neurons and neuroendocrine cells. Therefore, the question was whether VMAT2 activity in these cell fragments is also modulated by G proteins and what kind of G proteins are involved. In permeabilized platelets, the G protein activator GMpp-Np decreases V_{max} and increases K_m of VMAT2 activity (Höltje et al. 2003; Fig. 5). The main G-protein α-subunit in platelets is Gαq. Using platelets from Gαq knock-out mice (kindly provided by S. Offermanns) it turned out that VMAT2 activity is regulated by Gαq in platelets while G-protein-mediated VMAT2 regulation in brain SSV is unaffected (Höltje et al. 2003, Fig. 6). Thus, VMAT2 activity may be regulated by different G pro-

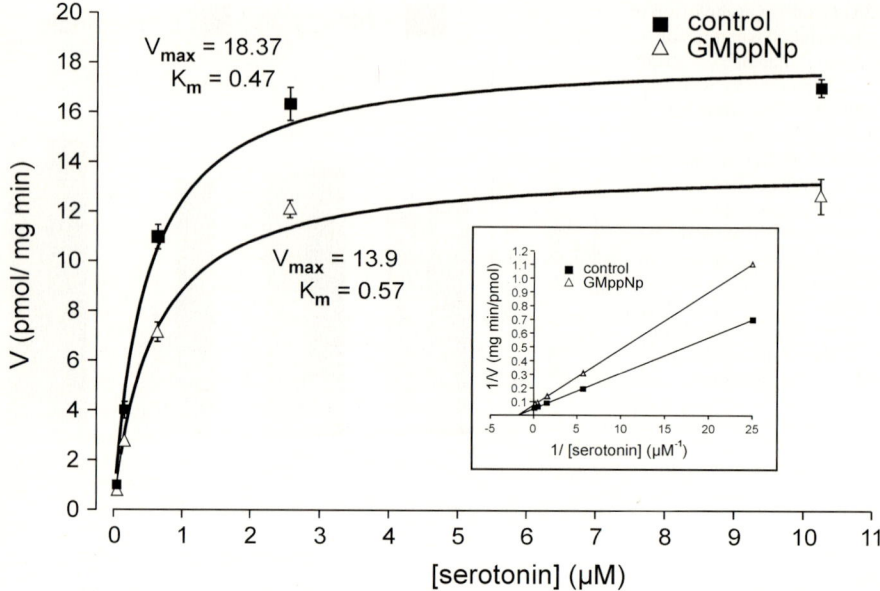

Fig. 5 Kinetic analysis of VMAT2 activity in platelets in the absence or presence of GMppNp. SLO-permeabilized platelets were loaded with increasing concentrations of [^{3}H] serotonin (*abscissa*) in the absence (*control*) or presence of 100 μM GMppNp. The G-protein activator decreased V_{max} and increased K_m. (From Höltje et al. 2003, with permission)

teins depending on the respective tissue. These data may also be taken as an indication that the interaction between the Gα-subunits and the transporters are not direct but rather mediated by cytosolic or vesicle attached proteins.

Glutamate uptake into SSV preparations from rat and mouse brain is also inhibited by the G-protein activator GMppNp (Pahner et al. 2003). In addition, both Gαo2 and Gαq localize to SSV of putative glutamatergic terminals (Pahner et al. 2003). However, GMppNp-mediated downregulation of glutamate uptake into SSV preparations is not affected in Gαq$^{-/-}$ as well as in Gα11$^{-/-}$ knock-out mice. Incubation of SSV with monoclonal antibodies specific for Gαo2, however, prevent the GMppNp-mediated inhibition of glutamate uptake, suggesting that Gαo2 may be also involved in the regulation of VGLUT activities (Winter et al., manuscript in preparation). Future studies are needed to find out whether the different VGLUTs, VGLUT1–3, are regulated by different G-proteins. Depending on the type of glutamatergic terminal, it may well be that VGLUTs exhibit a similar promiscuity for their regulation like VMAT2.

Putative upstream and downstream signals of vesicular G proteins

The presence of G-protein heterotrimers on secretory vesicles and the fact that vesicular properties in cell free preparations can be modulated by activating G proteins supports the idea that vesicular G proteins may be switched on from the luminal side. On the other side, G-protein α-subunits responsible for the transmitter transporter downregulation appear to

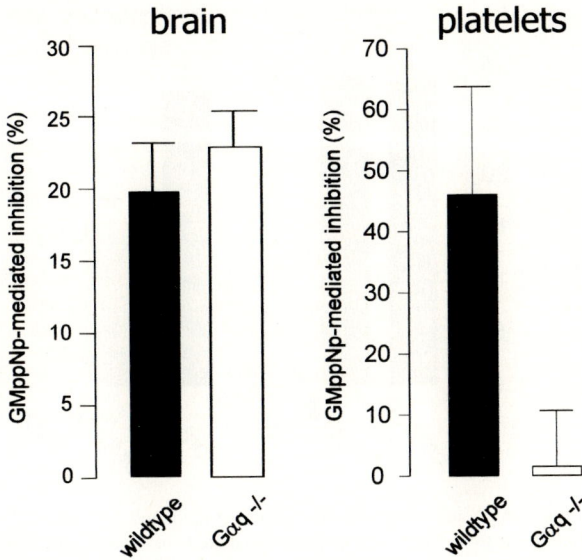

Fig. 6A, B Differential effects of Gαq on VMAT2 activity in brain and in platelets. **A** Crude synaptic vesicles (LP2), prepared from either wildtype or Gαq$^{-/-}$ mice were loaded with [^{3}H] serotonin in the presence of 100 μM GMppNp for 10 min. *Columns* represent the GMppNp-mediated inhibition of uptake expressed as percent of the uptake obtained in the absence of GMppNp. Data are combined from three individual experiments (mean±SD). **B** SLO permeabilized platelets from either wildtype or Gαq$^{-/-}$ mice were subjected to serotonin uptake for 15 min in the absence or presence of 100 μM GMppNp. *Columns* represent the GMppNp-mediated inhibition of uptake expressed as percent of the uptake obtained in the absence of GMppNp. Data are combined from five individual experiments (mean±SD). No inhibition was observed in Gαq$^{-/-}$ platelets. (From Winter et al. to be published and modified from Höltje et al. 2003, with permission)

require cytosolic signal cascades with elements that are not necessarily associated with the vesicles.

Vesicular filling as a regulator of vesicle-associated G proteins

As outlined above (see "Variability in the transmitter content of individual vesicles") vesicular filling can be varied by an overexpression of transmitter transporters or increasing the availability of transmitter. The opposite model characterized by empty vesicles is almost impossible to obtain by pharmacological means because all drugs interfere more or less with the vesicular transmitter transporters. However, empty but otherwise fully equipped vesicles are required to study the impact of vesicular filling on G-protein regulation. Platelets are perfectly equipped for monoamine uptake and storage and their VMAT2 activity is downregulated by Gαq. Platelets, however, are unable to synthesize serotonin, which they accumulate during their passage through the capillary system of gut villi. Using platelets from mice deficient for the peripheral tryptophan hydroxylase (Tph)1 (Walther et al. 2001), the relation between vesicular filling and G-protein-mediated vesicular transmitter transporter regulation has been addressed. In Tph 1$^{-/-}$ platelets which contained only 6% of the serotonin of wildtype (wt) and no other monoamines, GMppNp fails to downregulate VMAT2 activity. Inhibition can be reconstituted when preloading plate-

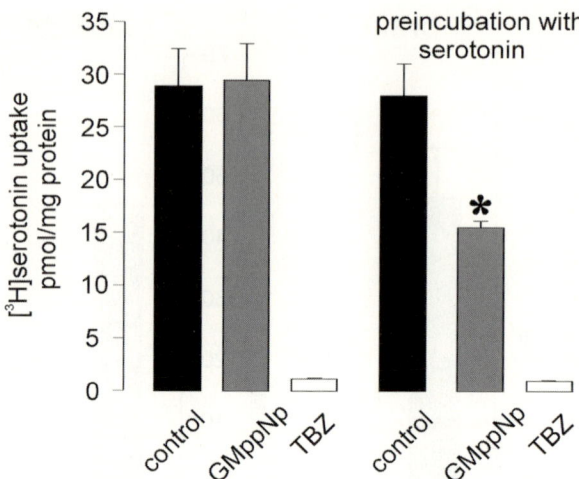

Fig. 7 The vesicular transmitter content activates the GMppNp-mediated downregulation of VMAT2 activity in platelets. Intact platelets from Tph1[-/-] mice were incubated for 30 min with buffer alone or supplemented with 15 μM nonlabeled serotonin. No GMppNp-induced inhibition was observed in platelets receiving just buffer during the preincubation. After preincubation and following SLO-permeabilization, the addition of GMppNp inhibited the vesicular [^{3}H] serotonin uptake by 50%. Values given in pmol/mg protein represent the mean of data obtained from three samples±SD. Statistical significance (p=0.022) was verified using Student's t test. (From Höltje et al. 2003, with permission)

lets, thereby adjusting serotonin levels to the one of wt platelets (Höltje et al. 2003; Fig. 7). These data provide the first evidence that the vesicular filling triggers G-protein activation and that this regulation is sustained by the vesicle itself. The data also suggest that a signal cascade can be switched on from the luminal side of an organelle. The nature of the involved putative receptor which might face the vesicular lumen is unknown. So far, all types of vesicles analyzed exhibit a G-protein regulation of their transmitter transporter activity. Therefore, one can tentatively speculate that the luminal sensor is provided by the transmitter transporters themselves which are specific for their respective transmitters independent from the type of the vesicle they reside on. This idea is supported by recent analysis from our lab using VMAT-transfected CHO cell lines which express Go-proteins but do not synthesize monoamines. Preloading permeabilized CHO-VMAT1 cells with dopamine noradrenaline, adrenaline, or serotonin increases the inhibition of the following [^{3}H] serotonin uptake by GMppNp (Brunk et al., manuscript in preparation).

Modulation of transporter activity by cytosolic components

While the vesicular filling may represent the upstream signal for G-protein activation, the question is how the G protein interferes with the transmitter transporter. So far, no indications have been obtained that this interaction is a direct one. In addition, using permeabilized cells effects of G-protein activators on transmitter uptake are more pronounced than when using isolated secretory vesicle preparation. This indicates that vesicles lose attached cytosolic components during purification which may be involved in the regulation of transmitter transporter activity by G proteins. Putative downstream signals could be cAMP-mediated processes. Increasing cAMP diminishes vesicular monoamine uptake in

PC 12 cells in a PKA-sensitive manner (Nakanishi et al. 1995a, b). Similar vesicular serotonin uptake mediated by VMAT2 (Pahner et al. 2003) but not the VMAT2-mediated serotonin uptake into platelets (S. Winter, M. Höltje, G. Ahnert-Hilger, unpublished observation) can be downregulated by cAMP. These observations suggest that, depending on the G protein involved (Gαo2 in brain, Gαq in platelets), different down stream signals work. When directly comparing vesicular uptake for serotonin or glutamate, cAMP had no effects on the latter, suggesting that different transmitter transporters are linked to different signal transduction pathways (Pahner et al. 2003).

Phosphorylation of VMAT2 has been shown to occur at serine residues by casein kinase II. However, modification of these residues do not change kinetic parameters. Whether phosphorylation is also involved in the GMppNp-mediated downregulation of VMAT2 is not clear so far and may be less likely given the fact that VMAT1 is not phosphorylated (Krantz et al. 1997). Protein kinase C-mediated phosphorylation may be involved in the regulation of VAChT- activity, since activators of PKC decrease binding of the VAChT inhibitor vesamicol (Clarizia et al. 1999).

Overexpression of Rab3A and Rab3B correlated with an increased amount of phosphoinositide 3-kinase have been shown to increase V_{max} of VMAT1 activity in PC 12 cells. These data indicate that small GTP-binding proteins might also be involved in the regulation of vesicular filling (Francis et al. 2002). However, it is not clear whether an increased expression of VMAT1 activity accounts for the increased vesicular transmitter uptake. Using rab3a knock-out mice, the GMppNp-mediated downregulation of serotonin uptake into SSV preparations from brain is unchanged compared to SSV preparations from wild-type animals (S. Winter, M. Höltje, G. Ahnert-Hilger, unpublished observation).

Uptake of glutamate and GABA into SSV can be inhibited by a cytosolic protein of 103 kDa named inhibitory protein factor (IPF; Özkan et al. 1997). This protein appears to be concentrated in synaptosomal cytosol and also interferes with glutamate, GABA ,and serotonin release probably due to an impaired uptake (Tamura et al. 2001). IPF could be one of the cytosolic factors involved in the G protein-mediated transmitter transporter downregulation.

Taken together, these data suggest that vesicular transmitter transporters may change their functional properties probably by involving cytosolic proteins and various types of phosphorylations. Whether and how these modulations are linked to heterotrimeric G proteins residing on secretory vesicles has yet to be determined.

Plasma membrane receptor-mediated changes of vesicular filling

Regulation of vesicular filling irrespective of whether the vesicles reside in neurons, neuroendocrine cells, or even platelets can be sustained at the vesicular level. Physiologically, such a regulation should be adapted to the surrounding environment, or, in the case of neurons, to the requirements of the neuronal network. Therefore, a vesicular regulation of transmitter content should be linked to plasma membrane receptors. There are only very few data showing a link between plasma membrane receptors and vesicular content. D2-like dopamine receptors reduce quantal size in PC 12 cells due to an inhibition of tryrosine hydroxylase activity (Pothos et al. 1998a). Treatment of midbrain dopamine neurons with glial-derived neurotrophic factor increases their quantal size (Pothos et al. 1998b). Thus, environmental changes of neurons and neuroendocrine cells can activate plasma membrane receptors, which then in turn influence transmitter filling mediated by vesicular G proteins.

Conclusions

Variations in the transmitter content of secretory vesicles are mediated by vesicle-associated heterotrimeric G proteins. $G\alpha o2$ and $G\alpha q$ regulate VMAT activities in brain or platelets, respectively; the G proteins involved in the control of VGLUTs have to be identified. As suggested from the platelet and other models, the vesicular content appears to activate the G protein. This signal transduction form the luminal site might be mediated by a vesicular G-protein coupled receptor, or, as an alternative possibility, by the transporter itself (Fig. 8). These novel functions of G proteins in the control of transmitter storage may link regulation of the vesicular content to intracellular signal cascades.

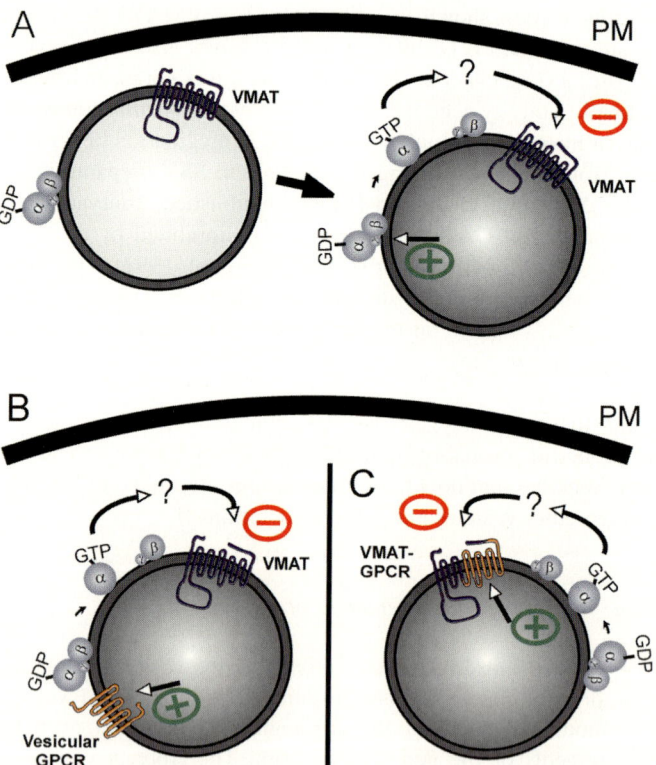

Fig. 8A, B Regulation of vesicular transmitter transporters by vesicle-associated G proteins. **A** As predicted for the VMAT regulation in platelets from Tph 1$^{-/-}$ mice, the transmitter content activates the G protein which then modifies transporter activity using intracellular signal cascades to be identified. **B** The structure sensing the transmitter content might be a G-protein-coupled receptor (*GPCR*) of unknown identity so far. As an alternative possibility, the transporter by its intravesicular loops senses the transmitter content (as a VMAT-GPCR) and turns the G protein on

References

Agnati LF, Zoli M, Strömberg I, Fuxe K (1995) Intracellular communication in the brain: wiring versus volume transmission. Neuroscience 69:711–726

Ahnert-Hilger G, Schäfer T, Spicher K, Grund C, Schultz G, Wiedenmann B (1994) Detection of G-protein heterotrimers on large dense core and small synaptic vesicles of neuroendocrine and neuronal cells. Eur J Cell Biol 65:26–38

Ahnert-Hilger G, Nürnberg B, Exner T, Schäfer T, Jahn R (1998) The heterotrimeric G protein Go2 regulates catecholamine uptake by secretory vesicles. EMBO J 17:406–413

Alfonso A, Grundahl K, Duerr JS, Han HP, Rand JB (1993) The *Caenorhabditis elegans unc-17* gene: a putative vesicular acetylcholine transporter. Science 261:617–619

Aronin N, DiFiglia M (1992) The subcellular localization of the G-protein Giα in the basal ganglia reveals its potential role in both signal transduction and vesicle trafficking. J Neurosci 12:3435–3444

Bai L, Xu H, Collins JF, Ghishan FK (2001) Molecular and functional analysis of a novel neuronal vesicular glutamate transporter. J Biol Chem 276:36764–36769

Barbour B, Häusser M (1997) Intersynaptic diffusion of neurotransmitter. Trends Neurosci 20:377–384

Bedet C, Isambert M-F, Henry J-P, Gasnier B (2000) Constitutive phosphorylation of the vesicular inhibitory amino acid transporter in rat central nervous system. J Neurochem 75:1654–1663

Bellocchio EE, Reimer RJ, Fremeau RT Jr, Edwards RH (2000) Uptake of glutamate into synaptic vesicle by an inorganic phosphate transporter. Science 289:957–960

Bruns D, Riedel D, Klingauf J, Jahn R (2000) Quantal release of serotonin. Neuron 28:205–220

Cases O, Seif I, Grimsby J, Gaspar P, Chen K, Pournin S, Muller U, Aguet M, Babinet C, Chen Shih J, De Maeyer E (1995) Aggressive bahaviour and altered amounts of brain serotonin and norepinephrine in mice lacking MAOA. Science 208:1763–1766

Chaudry FA, Reimer RJ, Bellocchio EE, Danbolt NC, Osen KK, Edwards RH, Storm-Mathisen J (1998) The vesicular GABA transporter, VGAT, localises to synaptic vesicles in sets of glycinergic as well as GABAergic neurons. J Neurosci 18:9733–9750

Clarizia AD, Gomez MV, Romano-Silva MA, Parsons SM, Prado VF, Prado MAM (1999) Control of binding of a vesamicol analog to the vesicular acetylcholine transporter. Neuroreport 10:2783–2787

Colliver TL, Pyott SJ, Achalabun M, Ewing AG (2000) VMAT-mediated changes in quantal size and vesicular volume. J Neurosci 20:5276–5282

Danbolt NC (2001) Glutamate uptake. Prog Neurobiol 65:1–105

Daniels GM, Amara S (1999) Regulated trafficking of the human dopamine transporter. Clathrin-mediated internalization and lysosomal degradation in response to phorbol esters. J Biol Chem 50:35794–35801

Deken SL, Beckmann ML, Boos L, Quick MW (2000) Transport rates of GABA transporters: regulation by the N-terminal domain and syntaxin 1A. Nature Neurosci 3:998–1003

Deken SL, Wang D, Qick MW (2003) Plasma membrane GABA transporters reside on distinct vesicles and undergo rapid regulated recycling. J Neurosci 23:1563–1568

Dhingra A, Jiang M, Wang T-L, Lyubarsky A, Savchenko A, Bar-Yehuda T, Sterling P, Birnbaumer L, Vardi N (2002) Light response of retinal ON bipolar cells requires a specific splice variant of Gαo. J Neurosci 22:4878–4884

Dumoulin A, Rostaing P, Bedet C, Levi S, Isambert M-F, Henry J-P, Triller A, Gasnier B (1999) Presence of the vesicular inhibitory amino acid transporter in GABAergic and glycinergic synaptic terminal boutons. J Cell Sci 112:811–823

Elmquist D, Quasterl DMJ (1965) A quantitative study of endplate potentials in isolated human muscle. J Physiol 178:505–529

Engel D, Pahner I, Schulze K, Frahm C, Jarry H, Ahnert-Hilger G, Draguhn A (2001) Plasticity of central inhibitory synapses through GABA metabolism. J Physiol (Lond) 535:473–485

Erickson JD, Eiden LE, Hoffman BJ (1992) Expression cloning of a reserpine-sensitive vesicular monoamine transporter. Proc Natl Acad Sci 89:10993–10997

Erickson JD, Varoqui H, Schäfer MK, Modi W, Diebler MF, Weihe E, Rand J, Eiden LE, Bonner TI, Usdin TB (1994) Functional identification of a vesicular acetylcholine transporter and its expression from a "cholinergic" gene locus. J Biol Chem 269:21929–21932

Erickson JD, Schäfer MK, Bonner TI, Eiden LE, Weihe E (1996) Distinct pharmacological properties and distribution in neurons and endocrine cells of two isoforms of the human vesicular monoamine transporter. Proc Natl Acad Sci 93:6166–5171

Francis SC, Sunshine C, Kirk KL (2002) Coordinate regulation of catecholamine uptake by rab3 and phosphoinositide 3-kinase. J Biol Chem 277:7816–7823

Fremeau RT Jr, Matthew DT, Pahner I, Nygaard GO, Tran CH, Reimer RJ, Bellocchio EE, Fortin D, Storm-Mathisen J, Edwards RH (2001) The expression of vesicular glutamate transporters defines two classes of excitatory synapse. Neuron 31:247–260

Fremeau RT, Burman J, Qureshi T, Tran CH, Proctor J, Johnson J, Zhang H, Sulzer D, Copenhagen DR, Storm-Mathisen J, Reimer RJ, Chaudhry FH, Edwards RH (2002) The identification of vesicular glutamate transporter 3 suggests novel modes of signaling by glutamate. Proc Natl Acad Sci 99:14488–14493

Frerking M, Wilson M (1996) Saturation of postsynaptic receptors at central synapses? Curr Opin Neurobiol 6:395–403

Fujiyama F, Furuta T, Kaneko T (2001) Immunocytochemical localization of candidates for vesicular glutamate transporterin the rat cerebral cortex. J Comp Neurol 435:379–387

Garzon M, Pickel VM (2000) Denritic and axonal targeting of the vesicular acetylcholine transporter to membranous cytoplasmic organelles in laterodorsal and pedunculopontine tegmental nuclei. J Comp Neurol 419:32–48

Geerlings A, Lopez-Corcuera B, Aragon C (2000) Characterization of the interactions between the glycine transporters GLYT1 and GLYT2 and the SNARE protein synatxin 1A. FEBS Lett 470:51–54

Gras C, Herzog E, Bellenchi GC, Bernard V, Ravassard P, Pohl M, Gasnier B, Giros B, El Mestikawy S (2002) A third vesicular glutamate transporter expressed by cholinergic and serotoninergic neurons. J Neurosci 22:5442–5451

Haase J, Killian A-M, Magnani F, Williams C (2001) Regulation of the serotonin transporter by interacting proteins. Biochem Soc Trans 29:722–728

Hajos N, Nusser Z, Rancz EA, Freund TF, Mody I (2000) Cell type- and synapse-specific variability in synaptic GABAA receptor occupancy. Eur J Neurosci 12:810–812

Hansson SR, Hoffmann BJ, Mezey E (1998) Ontogeny of vesicular monoamine transporter mRNAs VMAT1 and VMAT2. The developing rat central nervous system. Brain Res Dev Brain Res 110:135–158

Hayashi M, Otsuka M, Morimoto R, Hitota S, Yatsushiro S, Takeda J, Yamamoto A, Moriyama Y (2001) Differentiation-associated Na$^+$-dependent inorganic phosphate cotransporter (DNPI) is a vesicular glutamate transporter in endocrine glutamatergic systems. J Biol Chem 276:43400–43406

Hell JW, Maycox PR, Stadler H, Jahn R (1988) Uptake of GABA by rat brain synaptic vesicles isolated by a new procedure. EMBO J 7:3023–3029

Hell JW, Maycox PR, Jahn R (1990) Energy dependence and functional reconstitution of the γ-aminobutyric acid carrier from synaptic vesicles. J Biol Chem 265:2111–2117

Hisano S, Hoshi K, Ikeda Y, Maruyama D, Kanemoto M, Ichijo J, Kojima I, Takeda J, Nogami H (2000) Regional expression of a gene encoding a neuron-specific Na$^+$-dependent inorganic phosphate cotransporter (DNPI) in the rat forebrain. Mol Brain Res 83:34–43

Höltje M, von Jagow B, Pahner I, Lautenschlager M, Hörtnagl H, Nürnberg B, Jahn R, Ahnert-Hilger G (2000) The neuronal monoamine transporter VMAT2 is regulated by the trimeric GTPase Go2. J Neurosci 20:2131–2141

Höltje M, Winter S, Walther D, Pahner I, Hörtnagl H, Ottersen OP, Bader M, Ahnert-Hilger G (2003) The vesicular monoamine content regulates VMAT2 activity through Gaq in mouse platelets. Evidence for autoregulation of vesicular transmitter uptake. J Biol Chem 278:15850–15858

Ishikawa T, Sahara Y, Takahashi T (2002) A single packet of transmitter does not saturate postsynaptic glutatmate receptors. Neuron 34:613–621

Jahn R, Südhof TC (1999) Membrane fusion and exocytosis. Annu Rev Biochem. 68:863–911

Jiang M, Gold MS, Boulay G, Spicher K, Pexton M, Brabet P, Srinivasan Y, Rudolph U, Ellison G, Birnbaumer L (1998) Multiple neuorological abnormalities in mice deficient in the G protein Go. Proc Natl Acad Sci 95:3269–3274

Johnson RG Jr (1988) Accumulation of biological amines into chromaffin granules: A model for hormone and neurotransmitter transport. Physiol Rev 68:232–307

Kaneko T, Fujiyama F (2002) Complementary distribution of vesicular glutamate transporter in the central nervous system. Neurosci Res 42:243–250

Krantz DE, Peter D, Liu Y, Edwards RH (1997) Phosphorylation of a vesicular monoamine transporter by casein kinase II. J Biol Chem 272:6752–6759

Lesch KP, Gross J, Wolozin BL, Murphy DL, Riederer P (1993) Extensive sequence divergence between the human and rat brain vesicular monoamine transporter: possible molecular basis for species differences in the susceptibility to MPP$^+$. J Neural Transm 93:75–82

Lin RC, Scheller RH (2000) Mechanism of synaptic vesicle exocytosis. Annu Rev Cell Dev Biol 16:19–49

Liu Y, Peter A, Roghani A, Schuldiner S, Prive GG, Eisenberg D, Brecha N, Edwards R (1992) A cDNA that suppresses MPP$^+$ toxicity encodes a vesicular amine transporter. Cell 70:539–551

Liu Y, Schweitzer E, Nirenberg MJ, Pickel VM, Evans CJ, Edwards RH (1994) Preferential localization of a vesicular monoamine transporter to dense core vesicles in PC 12 cells. J Cell Biol 127:1419–1433

Mainen ZF, Malinow R, Svoboda K (1999) Synaptic calcium transients in single spines indicate that NMDA receptors are not saturated. Nature 399:151–155

Maycox PR, Hell JW, Jahn R (1990) Amino acid neurotransmission: spotlight on synaptic vesicles. Trends Neurosci 13:83–87

McIntire SL, Reimer RJ, Schiske K, Edwards RH, Jorgensen EM (1997) Identification of the vesicular GABA transporter. Nature 389:870–876

Mozhayeva MG, Sara Y, Liu X, Kavalali ET (2002) Development of vesicle pools during maturation of hippocampal synapses. J Neurosci 22:654–665

Nakanishi N, Onozawa S, Matsumoto R, Hasegawa H, Yamada S (1995a) Cyclic AMP-dependent modulation of vesicular monoamine transport in pheochromocytoma cells. J Neurochem 64:600–607

Nakanishi N, Onozawa S, Matsumoto R, Kurihara K, Ueha T, Hasegawa H, Minami N (1995b) Effects of protein kinase inhibitors and protein phosphatase inhibitors on cylic AMP-dependent downregulation of vesicular monoamine transport in pheochromocytoma PC12 cells. FEBS Lett 368:411–414

Ni B, Rostock PR Jr, Nadi NS, Paul SM (1994). Cloning and expression of a cDNA encoding a brain-specific Na$^+$-dependent inorganic phosphate cotransporter. Proc Natl Acad Sci 91:5607–5611

Nürnberg B, Ahnert-Hilger G (1996) Potential roles of heterotrimeric G proteins of the endomembrane system. FEBS Lett 389:61–65

Offermanns S (1999) New insights into the in vivo function of the heterotrimeric G-protein through gene deletion studies. Naunyn-Schmiedebergs Arch Pharmacol 360:5-13

Özkan ED, Lee FS, Ueda T (1997) A protein factor that inhibits ATP-dependent glutamate and γ-aminobutyric acid accumulation into synaptic vesicles: Purification and initial characterization. Proc Natl Acad Sci 94:4137–4142

Pahner I, Höltje M, Winter S, Nürnberg B, Ottersen OP, Ahnert-Hilger G (2002) Subunit composition and functional properties of G-protein heterotrimers on rat chromaffin granules. Eur J Cell Biol 81:449–456

Pahner I, Höltje M, Winter S, Takamori S, Bellocchio EE, Spicher K, Laake P, Nürnberg B, Ottersen OP, Ahnert-Hilger G (2003) Functional G-protein heterotrimers are associated with vesicles of putative glutamatergic terminals: implications for regulation of transmitter uptake. Mol Cell Neurosci 23:398–413

Peter D, Jimenez J, Liu Y, Kim J, Edwards RH (1994) The chromaffin granule and synaptic vesicle amine transporters differ in substrate recognition and sensitivity to inhibitors. J Biol Chem 269:7231–7237

Peter, D, Liu Y, Sternini, C, de Giorgio R, Brecha N, Edwards RH (1995) Differential expression of two vesicular monoamine transporters. J Neurosci 15:6179–6188

Pieribone VA, Shupliakov O, Brodin L, Hilfiker-Rothenfluh S, Czernik AJ, Greengard P (1995) Distinct pools of synaptic vesicles in neurotransmitter release. Nature 375:493–497

Pothos EN, Przedborski S, Davila V, Schmitz Y, Sulzer D (1998a) D2-like dopamine receptor reduces quantal size in PC12 cells. J Neurosci 18:5575–5585

Pothos EN, Davila V, Sulzer D (1998b) Presynaptic recording of quanta from midbrain dopamine neurons and modulation of quantal size. J Neurosci 18:4106–4118

Pothos EN, Larsen KE, Krantz DE, Liu Y-j, Haycock JW, Setlik W, Gershon MD, Edwards RH, Sulzer D (2000) Synaptic vesicle transporter expression regulates vesicle phenotype and quantal size. J Neurosci 20:7297–7306

Reimer RJ, Fon EA, Edwards RH (1998) Vesicular neurotransmitter transport and the presynaptic regulation of quantal size. Curr Opin Neurobiol 8:405–412

Reimer RJ, Fremeau RT Jr, Bellocchio EE, Edwards RH (2001) The essence of excitation. Curr Opin Cell Biol 13:417–421

Sagné C, El Mestikawy S, Isambert M-F, Hamon M, Henry J-P, Giros B, Gasnier B (1997) Cloning of a functional vesicular GABA and glycine transporter by screening of genome databases. FEBS Lett 417:177–183

Sakata-Haga H, Kanemoto M, Maruyama D, Hoshi K, Mogi K, Narita M, Okada N, Ikeda Y, Nogami H, Fukui Y, Kojima I, Takeda J, Hisano S (2001) Differential localization and colocalization of two neuron-types of sodium-dependent inorganic phosphate cotransporters in rat forebrain. Brain Res 902:142–155

Schäfer MKH, Varoqui H, Defamie N, Weihe E, Erickson JD (2002) Molecular cloning and functional identification of mouse vesicular glutamate transporter 3 and its expression in subsets of novel excitatory neurons. J Biol Chem 277:50734–50748

Schuldiner S, Shirvan A, Linial M (1995) Vesicular neurotransmitter transporters: From bacteria to humans. Physiol Rev 75:369–392

Schütz B, Schäfer MK, Eiden LE, Weihe E (1998) Vesicular amine transporter expression and isoforms selection in developing brain, peripheral nervous system and gut. Brain Res Dev Brain Res 106:181–204

Sondek J, Siderovski DO (2001) Gγ-like (GGL) domains: new frontier in G-protein signaling and β-propeller scaffolding. Biochem Pharmacol 61:1329–1337

Song H-j, Ming G-l, Fon E, Bellocchio E, Edwards E, Poo M-m (1997) Expression of a putative vesicular acetylcholine transporter facilitates quantal transmitter release. Neuron 18:815–826

Stevens CF, Wesseling JF (1999) Identification of a novel process limiting the rate of synaptic vesicle cycling at hippocampal synapses. Neuron 24:1017–1028

Sutton B, Fasshauer D, Jahn R, Brünger AT (1998) Crystal structure of a SNARE complex involved in synaptic vesicle exocytosis at 2.4 Å resolution. Nature 395:347–353

Takamori S, Rhee JS, Rosenmund C, Jahn R (2000a) Identification of a vesicular glutamate transporter that defines a glutamatergic phenotype in neurons. Nature 407:189–194

Takamori S, Riedel D, Jahn R (2000b) Immunoisolation of GABA-specific synaptic vesicles defines a functionally distinct subset of synaptic vesicles. J Neurosci 20:4904–4911

Takamori S, Rhee JS, Rosenmund C, Jahn R (2001) Identification of differentiation-associated brain-specific phosphate transporter as a second vesicular glutamate transporter (VGLUT2) J Neurosci 21:RC182

Takamori S, Malherbe P, Broger C, Jahn R (2002) Molecular cloning and functional characterization of human vesicular glutamate transporter 3. EMBO Rep 3:798–803

Tamura Y, Özkan ED, Bole DG, Ueda T (2001) IPF, a vesicular uptake inhibitor protein factor, can reduce the Ca^{2+}-dependent, evoked release of glutamate, GABA and serotonin. J Neurochem 76:1153–1164

Travis ER, Wang Y-M, Michael DJ, Caron MG, Wightman RM (2000) Differential quantal release of histamine and 5-hydroxytryptamine from mast cells of vesicular monoamine transporter 2 knock out mice. Proc Natl Acad Sci 97:162–167

Usdin TB, Eiden LE, Bonner TI, Erickson JD (1995) Molecular biology of the vesicular ACh transporter. Trends Neurosci 18:218–224

Van der Kloot W (1991) The regulation of quantal size. Prog Neurobiol 36:93–103

Van der Kloot W, Colasante C, Cameron R, Malgó J (2000) Recycling and refilling of transmitter quanta at the frog neuromuscular junction. J Physiol 523:247–258

Van der Kloot W, Malgo J, Cameron R, Colasante C (2002) Vesicle size and transmitter release at the frog neuromuscular junction when quantal acetylcholine content is increased or decreased. J Physiol 541:385–389

Varoqui H, Diebler MF, Meunier FM, Rand JB, Usdin TB, Bonner TI, Eiden LE, Erickson JD (1994) Cloning and expression of the vesamicol binding protein from the marine ray Torpedo. Homology with the putative vesicular acetylcholine transporter UNC-17 from Caenorhabditis elegans. FEBS Lett 342:97–102

Varoqui H, Erickson JD (1996) Active transport of acetylcholine by the human vesicular acetylcholine transporter. J Biol Chem 271:27229–27232

Vogel C, Mössner R, Gerlach M, Heinemann T, Murphy DL, Riderer P, Lesch K-P, Sommer C (2003) Absence of thermal hyperalgesia in serotonin transporter-deficient mice. J Neurosci 23:708–715

Walther DJ, Peter JU, Bashammakh S, Hörtnagl H, Voits M, Fink H, Bader M (2001) Synthesis of serotonin by a second tryptophan hydroxylase isoform. Science 299:76

Weihe E, Tao-Chen JH, Schäfer MKH, Erickson JD, Eiden LE (1996) Visualization of the vesicular acetylcholine transporter in cholinergic nerve terminals and its targeting to a specific population of small synaptic vesicles. Proc Natl Acad Sci 93:3547–3552

Williams J (1997) How does a vesicle know it is full? Neuron 18:683–686

Wolosker H, de Souza DO, de Meis L (1996) Regulation of glutamate transport into vesicles by chloride and proton gradient. J Biol Chem 271:11726–11731

Instructions for authors

1 Legal requirements

The author(s) guarantee(s) that the manuscript will not be published elsewhere in any language without the consent of the copyright holders, that the rights of third parties will not be violated, and that the publisher will not be held legally responsible should there be any claims for compensation.

Authors wishing to include figures or text passages that have already been published elsewhere are required to obtain permission from the copyright holder(s) and to include evidence that such permission has been granted when submitting their papers. Any material received without such evidence will be assumed to originate from the authors.

Manuscripts must be accompanied by the "Copyright Transfer Statement".

Please include at the end of the acknowledgements a declaration that the experiments comply with the current laws of the country in which they were performed.

2 Editorial procedure

Manuscripts should be submitted in English, together with one set of illustrations and a complete pdf file, to the editor in charge.

The author is responsible for the accuracy of the references.

3 Manuscript preparation

To help you prepare your manuscript, Springer offers a template that can be used with Winword 7 (Windows 95), Winword 6 and Word for Macintosh.

For details see point 4.

All manuscripts are subject to copy editing.

- **Title page**
  - The name(s) of the author(s)
  - A concise and informative title
  - The affiliation(s) and address(es) of the author(s)
  - The e-mail address, telephone and fax numbers of the communicating author

- **Abstract.** Each paper must be preceded by an abstract presenting the most important results and conclusions.

- **Abbreviations** should be defined at first mention in the abstract and again in the main body of the text and used consistently thereafter

A list of **symbols** should follow the abstract if such a list is needed. Symbols must be written clearly. The international system of units (SI units) should be used. The numbering of chapters should be in decimal form.

Footnotes on the title page are not given reference symbols. Footnotes to the text are numbered consecutively; those to tables should be indicated by superscript lower-case letters (or asterisks for significance values and other statistical data).

Acknowledgements. These should be as brief as possible. Any grant that requires acknowledgement should be mentioned. The names of funding organizations should be written in full.

Funding. Authors are expected to disclose any commercial or other associations that might pose a conflict of interest in connection with submitted material. All funding sources supporting the work and institutional or corporate affiliations of the authors should be acknowledged.

- **References**
The list of References should only include works that are cited in the text and that have been published or accepted for publication. Personal communications should only be mentioned in the text.

In the text, references should be cited by author and year (e.g. Hammer 1994; Hammer and Sjöqvist 1995; Hammer et al. 1993) and listed in alphabetical order in the reference list.

Examples:

Monographs:
Snider T, Grand L (1982) Air pollution by nitrogen oxides. Elsevier, Amsterdam

Anthologies and proceedings:
Noller C, Smith VR (1997) Ultraviolet selection pressure on earliest organisms. In: Kingston H, Fulling CP (eds) Natual environment background analysis. Oxford University Press, Oxford, pp 211–219

Journals:
Meltzoff AN, Moore MK (1977) Imitation of facial and manual gestures by human neonates. Science 198:75–78

If available the Digital Object Identifier (DOI) of the cited literature should be added at the end of the reference in question.

- **Illustrations and Tables**
All figures (photographs, graphs or diagrams) and tables should be cited in the text, and each numbered consecutively throughout. Figure parts should be identified by lower-case roman letters. The placement of figures and tables should be indicated in the left margin. For submission of figures in electronic form see below

Line drawings. Please submit good-quality prints. The inscriptions should be clearly legible.

Half-tone illustrations (black and white and color). Please submit well-contrasted photographic prints with the top indicated on the back. Magnification should be indicated by scale bars.

Figure legends must be brief, self-sufficient explanations of the illustrations. The legends should be placed at the end of the text.

Tables should have a title and a legend explaining any abbreviation used in that table. Footnotes to tables should be indicated by superscript lower-case letters (or asterisks) for significance values and other statistical data.

4 Electronic submission of final version

Please send only the final version of the article, as accepted by the editors.

Preparing your manuscript

The template is available:

→ via ftp:
  Address: ftp.springer.de/
  User ID: ftp
  Password: your own e-mail address
  – Directory: /pub/Word/journals
  – File names: either sv-journ.zip or
    sv-journ.doc and sv-journ.dot

→ via browser
  – http://www.springer.de/author/index.html

The zip file should be sent uuencoded.

Layout guidelines
1. Use a normal, plain font (e.g., Times Roman) for text.
   Other style options:
   – for textual emphasis use italic types.
   – for special purposes, such as for mathematical vectors, use boldface type.
2. Use the automatic page numbering function to number the pages.
3. Do not use field functions.
4. For indents use tab stops or other commands, not the space bar.
5. Use the table functions of your word processing program, not spreadsheets, to make tables.
6. Use the equation editor of your word processing program or MathType for equations.
7. Place any figure legends or tables at the end of the manuscript.
8. Submit all figures as separate files and do not integrate them within the text.

Data formats
Save your file in two different formats:

1. RTF (Rich Text Format) or Word compatible Word 95/97
2. pdf (a single pdf file including text, tables and figures)

Illustrations

The preferred figure formats are EPS for vector graphics exported from a drawing program and TIFF for halftone illustrations. EPS files must always contain a preview in TIFF of the figure. The file name (one file for each figure) should include the figure number. Figure legends should be included in the text and not in the figure file.

Scan resolution: Scanned line drawings should be digitized with a minimum resolution of 800 dpi relative to the final figure size. For digital halftones, 300 dpi is usually sufficient.

Color illustrations: Store color illustrations as RGB (8 bits per channel) in TIFF format.

General information on data delivery

Please send us a zip file (text and illustrations in separate files) either:

→ Via ftp.springer.de
  (to our ftp.server; log-in "anonymous"; password: your e-mail address; further information in the readme file on the server)

→ By e-mail
  (only suitable for small volumes of data)
→ or on any of the following media:
  – On a diskette [you may use .tar, .zip, .gzip (.gz), .sit, and compress (.Z)]
  – On a ZIP cartridge
  – On a CD-ROM

Please always supply the following information with your data: journal title, operating system, word processing program, drawing program, image processing program, compression program.

The file name should be memorable (e.g., author name), have no more than 8 characters, and include no accents or special symbols. Use only the extensions that the program assigns automatically.

5 Proofreading

Authors should make their proof corrections on a printout of the pdf file supplied, checking that the text is complete and that all figures and tables are included. After online publication, further changes can only be made in the form of an Erratum, which will be hyperlinked to the article. The author is entitled to formal corrections only. Substantial changes in content, e.g. new results, corrected values, title and authorship are not allowed without the approval of the editor in charge. In such a case please contact the Editor in charge before returning the proofs to the publisher.

6 Offprints, Free copy

You are entitled to receive a pdf file of your article for your own personal use. Orders for offprints can be placed by returning the order form with the corrected proofs. One complimentary copy of the issue in which your article appears is supplied.

Printing: Saladruck Berlin
Binding Lüderitz&Bauer, Berlin